THE BAMBOOS OF SABAH

THE BAMBOOS OF SABAH

Soejatmi Dransfield

Herbarium, Royal Botanic Gardens,
Kew, U.K.

in association with

Herbarium, Forest Research Centre,
Forestry Department,
Sabah, Malaysia

with illustrations by
Martin Molubin, Soejatmi Dransfield & Mary Watt

SABAH FOREST RECORDS, NO.14
FORESTRY DEPARTMENT, SABAH, MALAYSIA
1992

First published 1992

The Bamboos of Sabah
ISBN 983-9554-03-4

Sabah Forest Records, No. 14
ISSN 0128-6471

General editor of Sabah Forest Records: Robert C. Ong
Editor of this number: K.M. Wong

CONTENTS

CONTENTS (continued)

FOREWORD

In 1958 Professor R.E. Holttum, one of the foremost bamboo specialists of our time, wrote "... the first thing a user of bamboo needs is the ability to distinguish one kind from another; and the second thing is the need of names which have a definite meaning." *The Bamboos of Sabah* is a culmination of years of work by Dr Soejatmi Dransfield that fulfils those needs. When she first looked at Sabah bamboos in the 1970s a great many species were still not documented and many had not even been systematically collected. As she became more convinced that a careful, practical guide was required, the Forestry Department in Sabah encouraged her to undertake field work and herbarium documentation in association with the Herbarium of the Forest Research Centre at Sepilok, Sandakan.

In this book Dr Soejatmi Dransfield emphasises the value of understanding bamboos through field studies. It is extremely gratifying to see her work supported by both the Royal Botanic Gardens, Kew, U.K., where she is an honorary research associate, and various agencies in Sabah which have common links with the Forestry Department in its endeavour to better understand and manage the plant resources of Sabah. In particular, at a time when many of us realize the extent of the importance of bamboo as a component of natural and modified forest communities, and as a useful group of plants, her book testifies to the importance of careful scientific work and spirited cooperation between institutions.

This book is the first practical guide to the bamboos of Sabah written by a specialist with first-hand experience working on Sabah bamboos. More than being a statement of fact, this embodies the dedication that Dr Soejatmi Dransfield has given as a taxonomist, persevering through her research in Sabah and back at Kew, to the time when the account was finalised and edited at the Forest Research Centre at Sepilok. Through her knowledge and hard work, we can now claim that the bamboos of Sabah have received proper taxonomic documentation.

Datuk Miller Munang
Director,
Forestry Department,
Sabah, Malaysia

PREFACE

The completion of this bamboo account has been delayed for many years, because of continued findings of new species in the state. In bamboo taxonomy complete specimens are often essential for correct naming. Usually specimens consist of a flowering branch and a leafy branch, or in many cases they consist of vegetative parts without flowers, and this has been the case with work on Sabah bamboos. Fieldwork is essential in bamboo taxonomy so as to be able to study the plants in the field, and frequent visits are often very useful. I was able to conduct fieldwork twice in Sabah, and on my second visit (1986) managed to collect material of a new genus later named *Sphaerobambos*, and of a new species of *Dinochloa*, of which incomplete specimens had been collected some years earlier. It is a mistake to generalize that bamboos are all widely distributed. It seems there are actually only a few species widely distributed. Most species have limited distribution, and are often endemic to relatively small areas. Many of the bamboo species occurring in Sabah are found only over a limited area. As I have not visited all areas of Sabah, I may well have missed species. Therefore this account must be regarded as a reflection of what is known of Sabah bamboos to date. I believe there will be more undescribed taxa waiting to be collected and new records to be made in Sabah.

Soejatmi Dransfield
Kew, 1 April 1992

ACKNOWLEDGEMENTS

The present account is the result of a survey conducted in 1979 and fieldwork in 1986. I would like to thank SAFODA for sponsoring the survey in 1979 and the Royal Botanic Gardens, Kew (UK), for financing the trip in 1986. I am grateful to the Sabah Forestry Department for logistic support. Most of the work has been carried out in the Herbarium, Royal Botanic Gardens, Kew, and I am most grateful for the working facilities there. I would also like to thank Mr Lee Ying Fah, Forest Research Centre, Sepilok, Sandakan, for encouraging me to write this bamboo account, Mr Wong Khoon Meng for reading the manuscript critically and giving constructive suggestions, the late Mr A. J. Hepburn for organizing the initial surveys so efficiently, and Mr Dewol Sundaling for helping in the field. Finally I am grateful to Datuk Miller Munang, the Director of the Sabah Forestry Department for supporting the publication of the account.

Thanks are due to the Editors of the Kew Bulletin and Sandakania for permission to reproduce a number of figures, Ms Mary Watt of Kew for preparing one illustration and Mr Martin Molubin of Sandakan for preparing the rest of the illustrations. The book was typeset by Miss Pung Vui Lee of the Forest Research Centre, Sepilok, Sandakan.

INTRODUCTION

Bamboos are of great importance in Tropical Asia and perhaps nowhere else are they used so intensively for such a wide range of purposes. They have always been associated with rural life. In Borneo where timber is regarded as the most important forest product, bamboos have been considered rather unimportant. They have traditionally been excluded from forest inventory because they are regarded as village plants, or found growing in logged-over or secondary forests, or wastelands. After forest fires or logging in Sabah climbing bamboos often become abundant and there is considerable concern that their growth may prevent the regeneration of forest trees (Liew 1973).

Bamboos are members of the Gramineae (the grass family) and form the tribe Bambuseae of the subfamily Bambusoideae. There are 49-59 genera of bamboos in the world and the number of species is estimated at between 800 and 1000. They occur in the tropical, subtropical and temperate regions of all continents, except Europe and Western Asia, between 46° north and 47° south, from the lowlands up to 4000 m. The majority, however, occur at low to medium elevation in the tropics, growing wild or are naturalized or cultivated, in a great variety of habitats. Their present-day distribution has been greatly influenced by man's activities. So far there are nine genera documented in Sabah, Malaysia, all native or naturalised except *Thyrsostachys* (with one species, *T. siamensis*) which is introduced from Thailand. The other genera are *Bambusa* (six species), *Dendrocalamus* (one species), *Gigantochloa* (two species), *Dinochloa* (nine species), *Schizostachyum* (seven species), *Racemobambos* (six species), *Yushania* (one species), and *Sphaerobambos* (one species). An additional unnamed genus is found in Sabah, represented by an undescribed species mentioned at the end of the account.

MORPHOLOGY

Bamboos can be distinguished from other grasses by 1) woody culms or stems, 2) complex aerial branching systems, 3) complex underground branching systems, 4) stalked leaf blades, 5) specialized sheaths on the young stem, 6) the usual presence of 6 (in some genera 3) stamens and 3 lodicules, and 7) a different (specialized) leaf anatomy.

To many people most bamboos look alike and often are not individually recognized, because they share several similar features. A bamboo plant consists of a segmented underground part (the rhizome) and a segmented above-ground part or culm. There is no central trunk as in trees. Holttum (1958) gave an excellent account on the morphology of the bamboo plant based on Malaysian bamboos. The description of bamboo morphology below is based on the accounts by Holttum (1958) and McClure (1966), and also on first-hand observation.

Rhizomes

The rhizomes form a complex subterranean system which becomes the foundation of the bamboo plant. Each individual rhizome is segmented. The basal part of the rhizome is narrower and slenderer than the upper part or distal part and the segments are shorter. The basal part is termed the *neck*. The upper part is much larger and the segments or internodes are larger. Each segment or internode of the upper part bears a bud, roots and root primordia. The apex will continue to grow and becomes a culm which emerges above ground. The buds in the rhizome segments will grow and develop into rhizomes as well.

There are two basic types of rhizome: 1) pachymorph rhizomes (together forming a sympodium), characteristic of many subtropical and tropical bamboos, including all the bamboos known in Sabah and 2) leptomorph rhizomes (each being a monopodium), found commonly in temperate bamboos.

Young shoots

The new growth of the rhizome apex into a young culm is called a young shoot (e.g., Figs. 2A, 3A) or *rebung* in Malay and Indonesian. It consists of short soft young internodes protected by sheaths. This shoot usually grows very slowly

2

at first but later rapidly elongates to become a mature culm. The *rebung* of most bamboo species are edible, but those of some species need a pretreatment before cooking in order to get rid of cyanide compounds which may be present.

The young shoot (at a stage when it shows well-developed sheaths) is often very useful in recognising bamboo species. Some species have young shoots with sheaths covered by irritant white, black or dark brown hairs, or *miang* (e.g., Fig. 4A); some have young shoots covered with white wax. Many features which are specific characters can be seen in young shoots or young culms. A synoptic key to genera and species based on young shoots in the fresh state is presented below:

Young shoots covered with conspicuous white wax, usually glabrous
Dinochloa

Young shoots without conspicuous white wax, usually hairy

    Young shoots usually with black or dark brown hairs
Bambusa
most *Gigantochloa* species

    Young shoots usually with brown to pale brown or white hairs
Schizostachyum
G. balui
Sphaerobambos hirsuta
Dendrocalamus asper
Thyrsostachys siamensis

    Young shoots glabrous
Racemobambos
Yushania

Culms

The culms of bamboo species found in Sabah may be erect, erect with pendulous or drooping tips, scrambling or climbing. They can be straight, slightly zig-zag or markedly zig-zag. The internodes are hollow with thin or thick walls, or solid. When young the internodes are usually rough, glabrous or hairy, and become smooth and glabrous with age. The nodes can be swollen in some species. In some bamboo species, such as *Dendrocalamus asper*, *Gigantochloa levis*, or *Bambusa vulgaris*, the lowermost nodes each bear a ring of adventitious or aerial roots.

The size of a culm, i.e. its diameter, is already determined earlier by the size or diameter of the young shoot. The diameter gradually decreases along a culm towards the apex. Midculm internodes are generally longer than the lowermost or uppermost ones. Each culm node bears a branch bud or primordium in the axil of its sheath, usually just above the sheath scar. The buds are arranged alternately in two rows along the culm.

The culm is usually not a good feature for recognising species. However, the culm of *Bambusa vulgaris* can be easily recognised; it is not entirely straight, but slightly sinuous and the nodes are not horizontal. In *Dinochloa* the culms are unlike those of other genera in Southeast Asia in that the culms, which are often zig-zag, twine around their supports.

Culm sheaths

The young shoots and young culms are protected by sheaths, called culm sheaths (or sometimes culm leaves), or *pelepah buluh* in Malay/Indonesian (e.g., Figs. 1A, 2A & B). These are often regarded as modified leaves. This culm sheath embraces the developing internode and will fall off when the culm becomes mature. A culm sheath consists of a sheath proper, a blade, a ligule, and sometimes two auricles. The blade attaches to the sheath along the narrow top part of the sheath (Fig. 1A). When it is old the blade usually will break away, except for example in *Bambusa multiplex*, whose blade is the continuation of the sheath (Figs. 3A, B). The ligule is an upgrowth from the top of the sheath, and the auricles are the lateral extensions on each side of the blade base (Figs. 1A, 4B). The structure of the culm sheath is very important for recognising species. In *Dinochloa* the sheath proper has a basal part which is rugose (e.g., Fig. 8B).

When young the sheaths are usually covered with irritant hairs, or *miang*, which can be white, pale brown, dark brown, golden brown or black. The sheaths also vary in colour, and can be green, yellowish green, pale green, bluish or purplish green, or yellow tinged with orange. The blades also vary with the species, and can be lanceolate (Fig. 6B) or broadly triangular (Fig. 4B), erect (Fig. 2A) or deflexed (Fig. 19B), and glabrous or hairy. The ligule can be short or long and sometimes fringed with long bristles (Fig. 14B) or teeth (Fig. 1A).

Branches

In many *Bambusa* species the branch buds at culm nodes will develop right from the lowermost node up to the uppermost ones. The primary branch is dominant and elongates, producing lateral branches at each node (Fig. 1C). In many other bamboo species only branch buds from the midculm nodes upwards will develop.

The branch system is often very useful for recognising a bamboo genus. In *Schizostachyum*, for example, the primary branch is not dominant, so that the branch complement at each node consists of several short branches of similar size (Figs. 23D, 24D). In *Dinochloa* the primary branch axis of the branch bud is dormant but produces from its basal nodes short secondary branches (Fig. 5C). In many *Racemobambos* species the primary branch axis is dominant with

4

small, short secondary branches from its basal nodes (Figs. 15A, 16B), and also produces many short branches at each node. The development of the branch complement in many species of bamboo has never been examined in detail.

Leaf blades and leaf sheaths

The ultimate branchlets each bear 8-18 leaf blades. Each leaf blade is connected to a leaf sheath which clasps the internode and the top of this sheath often has auricles (e.g., Fig. 1E). The leaf blade differs from the blade of a culm sheath by having a stalk or petiole. In most cases the leaf blade does not yield good characters for recognising species.

Inflorescences and flowering behaviour

Because the famous Chinese Giant Panda depends on bamboos for its food, most people are familiar with one of the most spectacular flowering processes in the Plant Kingdom. It has long been known in China and India that many bamboos flower at the same time, at intervals of 20-120 years, and then die, to be replaced by seedlings. It is a mistake, however, to assume all bamboos flower in this way as there are at least three main flowering types: 1) gregarious: a whole bamboo population flowers over a period of 2-3 years and then dies, though the rhizomes may be still alive; 2) sporadic: individuals flower seasonally or occasionally, and only the flowering culms will die afterwards, while the rhizomes continue to live; 3) continuous: individuals produce flowers all the year round, where the culms which have produced flowers will die, and will be replaced by new culms all the time, because the rhizomes continue to live and grow. These three types of flowering behaviour can be observed in bamboos found in Sabah. *Racemobambos gibbsiae* and *R. hepburnii* have a gregarious flowering habit (Campbell 1985; Wong *et al.* 1988); sporadic flowering can be found in most *Dinochloa* species, and the continuous flowering habit is characteristic of the majority of *Schizostachyum* species.

As members of the grass family, bamboos have typically compound inflorescence (or flowering branches) consisting of many flowers. Bamboo flowers, like those of other grasses, are usually very small and usually pass unnoticed; they are termed *florets*, each comprises a lemma, a palea, 3 lodicules (sometimes absent), 3 or 6 stamens, and an ovary with 3 stigmas. Usually two or more florets (rarely just one) are borne along jointed branchlets together with one to several modified basal sheaths or glumes. The whole structure is called a *spikelet* (Fig. 16D), arranged in a flowering branch or an inflorescence (Fig. 16B). There are two basic structures of inflorescence in bamboo. The first is the *semelauctant* type, a term introduced by McClure (1966), to indicate a determinate inflorescence in bamboo, in which the spikelets are borne in a raceme or a simple panicle, and emerge almost simultaneously, and then die almost simultaneously. The basic unit in the semelauctant inflorescence is a spikelet, just as in other grasses. The genus *Racemobambos* and *Yushania tessellata* possess semelauctant inflorescences. The second type is the *iterauctant* type, defined also by McClure for indicating the indeterminate nature of some bamboo inflorescences. In this type of inflorescence the basic unit is called a pseudospikelet, the distal portion of which resembles a true spikelet and the basal portion bearing buds, each of which is supported by a sheath and another bract-like structure called a prophyll. These buds will eventually develop into pseudospikelets (Fig. 14G), the process continuing almost indefinitely until the culm's reserves are exhausted. Iterauctant inflorescences can be borne terminating leafy branches (such as in *Schizostachyum*) (Fig. 23B), or on short lateral branches (such as in *Gigantochloa*) (Fig. 13E), or on long leafless branches of up to 3 m long (such as in *Dinochloa*) (partly shown in Figs. 10D & 11C).

Fruits

The fruit in the grass family is typically a grain or a caryopsis comprising a pericarp enclosing the seed; the seed itself consists of an endosperm (a food storage organ), a small embryo, which bears a radicle, a plumule and a scutellum. The scutellum is an organ absorbing or transmitting food during the process of germination. In bamboos, caryopsis structure varies with the genera. In *Bambusa*, *Gigantochloa*, *Racemobambos* and some *Dendrocalamus* species, the caryopsis has a relatively thin pericarp enclosing the seed, just as in other grasses. In *Schizostachyum* the caryopsis has a thick and hard pericarp easily separated from the seed. In these two types of caryopsis the embryo is basally situated. In *Dinochloa* and *Sphaerobambos* (Fig. 29D) the fruit is a berry, which is usually globose and has a thick and fleshy pericarp enclosing a large embryo; the endosperm is much reduced forming a thin layer between the embryo and the pericarp. In *Sphaerobambos* the embryo is basally situated, whereas in *Dinochloa* it is lateral. Bamboo fruits are usually not regarded as helpful for recognising species, but in *Dinochloa* the fruit is one of the most important characters for differentiating the species (compare Figs. 5H, 6H, 7K).

A KEY TO THE GENERA

1.   Inflorescences semelauctant, in the form of a raceme or a simple panicle, with a spikelet as the basic unit 2
    Inflorescences iterauctant, in the form of a compound flowering branch, with a pseudospikelet as the basic unit 3

2.   Scrambling bamboos; branch complement of few to many branches with the primary branch dominant; stamens 6 .... *Racemobambos*
    Erect bamboo; branch complement of several (usually 5) branches, all short and of similar size; stamens 3 .... *Yushania*
                                                                (*Y. tessellata*)

3.   Climbing or scrambling bamboos; fruit a berry, with fleshy pericarp and much reduced endosperm, but large embryo 4
    Erect bamboos, or erect then leaning bamboos; fruit a caryopsis with hard, thin or relatively thick pericarp, small embryo and well developed endosperm 5

4.   Climbing bamboos; culm zig-zag; the base of the culm sheath prominent, rugose; spikelet with only one floret; embryo laterally situated .... *Dinochloa*
    Erect, then scrambling bamboo; culm straight or slightly zig-zag; base of the culm sheath not prominent; spikelet comprising 3-5 florets; embryo basally situated .... *Sphaerobambos*

5.   Culms erect with drooping tips, with thin walls; branch complement of many branches at each node, all of similar size 6
    Culms erect, rarely with drooping tips, with relatively thick walls; branches few at each node, primary branch usually dominant 7

6.   Culm sheaths usually covered with white to light brown (rarely dark brown) hairs .... *Schizostachyum*
    Culm sheaths covered with black hairs ....
    .... "Bambusa" sp. related to *B. wrayi*
                                                                   (see end of account)

7.   Culms about 3 cm in diameter, with relatively thick walls and a small lumen; leaf blades very narrow (typically less than 0.8 cm) ....
    .... *Thyrsostachys*
                                                                       (*T. siamensis*)

Culms over 4 cm in diameter, with moderately thick walls; leaf blades broader 8

8.   Culm sheaths usually covered with dark hairs, blades usually erect .... *Bambusa*
Culm sheaths usually covered with white or light brown to dark brown hairs, blades spreading or deflexed 9

9.   Culms with short basal internodes covered with velvety golden brown hairs; nodes swollen, bearing aerial roots; young shoots covered with dark hairs .... *Dendrocalamus*
(*D. asper*)
Culms with longer basal internodes with scattered white or dark brown hairs; nodes not swollen, occasionally bearing aerial roots; young shoots covered with pale hairs to dark brown hairs .... *Gigantochloa*

BAMBUSA Schreber

(Malay: *bambu*; latinized form of vernacular name)

Schreber, Gen. Plant. ed. 8, 1: 236 (1789).

Closely tufted bamboos. Culms erect, hollow, with relatively thick walls, usually glabrous. Branch complement of several branches with the middle branch dominant and producing branches at each node. Culm sheaths usually covered with dark hairs, usually with well-developed auricles; blades erect, usually triangular. Inflorescence borne on branches of leafless culms. Pseudospikelets clustered at the nodes of the inflorescence axis, in general laterally flattened; spikelets comprising several glumes, several florets, rachilla internodes long and slender; lemma glabrous; palea 2-keeled, glabrous; lodicules 2-3, fringed with hairs; stamens 6, filaments free, anthers with or without apiculate tips; ovary with a conspicuous style, and 3 stigmas.

Distribution. The genus *Bambusa* is found throughout tropical Asia, comprising about 37 species. One of the species, *B. vulgaris*, is found all over the tropics, planted or naturalized in wastelands or river banks. There are at least six species in Sabah; they are found in cultivation, but often also escape from cultivation and become naturalised. In the following key to identification of species, *B. bambos*, a useful and common species in mainland South East Asia, has been included although it has yet to be recorded for Sabah.

Habitat. Species of *Bambusa* are usually found growing in open areas in the lowlands or on hill sides, on various types of soil, but are found abundantly in moist places like river banks.

Uses. In Sabah, where there are several good-quality native bamboos, *Bambusa* species are rarely utilized with the exception of *B. vulgaris*, of which the culms are used for poles.

Notes. The genus can be recognized by its thick-walled culms, dominant primary branch with several secondary branches, and erect triangular blades of the culm sheaths. In Sabah a small bamboo plant with variegated leaf blades is sometimes planted as an ornamental. This bamboo is also planted in Peninsular Malaysia and was included by Holttum (1958) under *Bambusa*. The flowers have never been recorded for this bamboo and its placement in *Bambusa* is not certain. For this reason this species is not included in this account.

Key to the species

1.   Tall bamboo, diameter 4-10 cm at base 2
      Medium-size or small bamboo, diameter not more than 4 cm 4

2.   Culms dull green, with thorny branches at lower nodes, old clumps
      with an impregnable tangle of spiny branches at the base 3
      Culms dark green, yellow or yellow with green stripes, with non-thorny
      branches at the base, old clumps without spiny branches; branches of
      upper nodes alternate clearly, forming a fan-like structure especially in
      young culms; planted or naturalized along rivers or roadsides ....
      .... *B. vulgaris*

3.   Young culms greyish green, young culm sheaths glabrous, orange
      with green stripes; ligule of culm sheath irregularly toothed with fringe
      of fine hairs along the edge; auricles with dense, dark velvety hairs;
      blades erect, covered with black hairs (not yet recorded for Sabah, but
      possibly also planted) .... *B. bambos*
      Young culms green; culm sheaths yellowish green with scattered dark
      brown hairs; ligule of culm sheath with bristles along the edge;
      auricles with scattered brown hairs on outer surface, and long dense
      bristles along the edge; blades erect first then spreading, with
      scattered dark brown hairs (not uncommon in Sabah, planted or
      naturalised) .... *B. blumeana*

4.   Culm sheaths covered with black hairs, blades broadly triangular,
      auricles prominent with long bristles; culm green often with yellow
      streaks (planted, rare in Sabah, introduced from Peninsular
      Malaysia) .... *B. heterostachya*
      Culm sheaths without black hairs, blades narrowly triangular,
      auricles small with or without bristles; culm green 5

5.   Culm sheaths rather rigid, blades separated or easily detached from
      culm sheath when dry; internodes often swollen in dwarf plants
      (planted, rare in Sabah) .... *B. tuldoides*
      Culm sheaths papery, blades remain attached with the sheath when dry;
      internodes not swollen 6

6.   Culms erect, about 8-10 m tall (rare, introduced, planted individual
      clumps) .... *Bambusa* sp.
      Culms erect first then drooping, 2 m tall (as a hedge) or 6 m long
      (planted, usually as hedges, sometimes as individual clumps) ....
      .... *B. multiplex*

1. **Bambusa blumeana** Schult.   (Fig. 1)

(named after C.L. Blume, a Dutch botanist)

Schultes in Syst. Veg. 7: 1343 (1830); Holttum in Gard. Bull. Sing. 16: 57 (1958).

Synonym:
Bambusa spinosa Roxb. in Hort. Beng. 25 (1814).

Culms up to 20 m tall, diameter 10 cm, erect, internodes dark green, slightly white waxy; branches borne at and spreading horizontally from the lowermost node upwards, bearing stout curved spines, middle branches branched again; young shoots with yellowish green sheaths and blades. Culm sheaths 18-32 cm long, 13-32 cm wide at the base, narrower near the apex, covered with appressed dark brown to black hairs, light green when young, becoming stramineous; blades narrowly triangular, tapering to a stiffly acute apex, erect first, then spreading, densely hairy abaxially when young, glabrescent with age, with dark hairs adaxially, base sloping and joined laterally to the stiff auricles; auricles with curly hairs on the surface, and bristles along the edge; ligule irregularly toothed, fringed with short dark brown bristles. Leaf blades 11-16 x 1.5-1.9 cm, base truncate, glabrous; sheaths glabrous; auricles not prominent, often with short bristles; ligule very short. Inflorescence consisting of spikelets groups, borne on branches of a leafless culm. Spikelets up to 5 cm long, comprising 5-12 florets, rachilla internodes hairy; lemma glabrous; palea as long as the lemma, 2-keeled, keels fringed.

Distribution. Native in Malesia, wild, naturalised or cultivated. In Sabah *B. blumeana* is found planted by rivers in villages in Tambunan and Keningau, also found growing naturalized by the field in Tenom.

Local names. *Tongkungon* (Kadazan, Dusun), *kawayen* (Murut), *bambu/buluh duri* (Malay).

Uses. In Sabah this species is rarely utilised, because it is not very common.

Notes. *B. blumeana* is related to *B. bambos*, a species native to the drier region of the mainland Asia and also called *bambu duri* in Malay.

2. **Bambusa heterostachya** (Gamble) Holttum   (Fig. 2)

(Greek, with flowers or spikes or spikelets of different sizes)

Holttum in Journ. Arn. Arb. 27: 341 (1946), Gard. Bull. Sing. 16: 65 (1958).

Synonyms:
Gigantochloa heterostachya Munro in Trans. Linn. Soc. 26: 125 (1868); Gamble in Ann. Roy. Bot. Gard. Calc. 7: 66 (1898).
Gigantochloa latispiculata Gamble, l. c. 67.

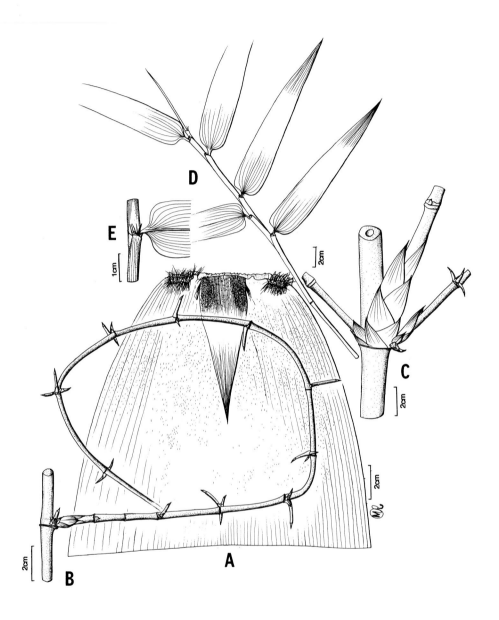

Fig. 1. *Bambusa blumeana*. A, Culm sheath. B, Basal spiny branch. C, Mid-culm branch complement. D, Leafy branch. E, Leaf sheath, detail. (From fresh material.)

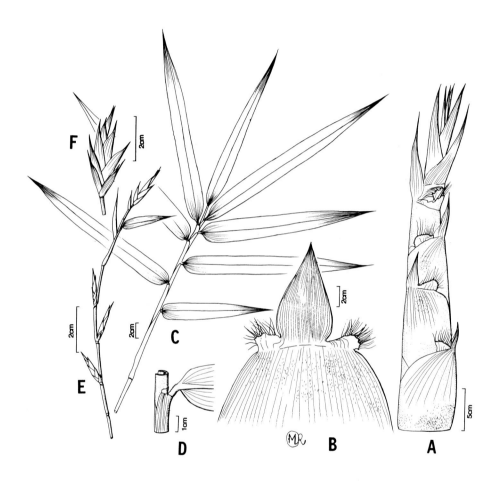

Fig. 2. *Bambusa heterostachya*. A, Culm shoot. B, Culm sheath. C, Leafy branch. D, Leaf sheath, detail. E, Flowering branch. F, Pseudospikelet. (A, B, E, F from SD 766; C, D from SD 769.)

13

Open tufted bamboo; culms about 7 m tall, diameter 3-6 cm at the base, erect, straight, internodes 30-80 cm long, with dark hairs below nodes when young, glabrous by age, nodes not swollen; young shoot purplish green. Branch complement several at each node. Culm sheaths green, covered with dark or black hairs; blades erect, light green when young, narrowly to broadly triangular, detached from the sheath at maturity, with scattered dark brown hairs; auricles large, with few, easily detached, curly bistles along the edge; ligule very short, entire. Leaf blades 24-40 x 2-4 cm, glabrous; sheaths glabrous or with black hairs; ligule very short. Inflorescences borne on leafless of leafy branches. Spikelets few at each inflorescence node, up to 3 cm long, comprising 2 glumes and 5-8 florets; glumes glabrous; lemma up to 20 mm long, with pointed tip, glabrous; palea shorter than the lemma, glabrous; lodicules 3.

Distribution. So far this species is only widely known in cultivation in Peninsular Malaysia; it is introduced in Sabah, planted on the grounds of the Ulu Dusun Agricultural Station and in the Pamol Oil Palm Plantation in Beluran.

Vernacular names. *Buluh telang, buluh galah* (Malay).

Uses. This species is introduced to Sabah for use in pollinating the flowers and harvesting the fruits of oil palms.

Notes. The straight, long, and relatively small-diameter culms are suitable for poles.

3. **Bambusa multiplex** (Loureiro) Raeuschel ex J.A. & J.H. Schult.   (Fig. 3)

(Latin, many branches)

Schultes in Roemer & Schultes, Syst. Veg. 1350-1351 (1830); Soderstrom & Ellis, Smiths. Contrib. Bot. 72: 36 (1988).

Synonyms:
Arundo multiplex Lour., Fl. Cochinch. 1: 58 (1790).
Ludolfia glaucescens Willd., Ges. Naturf. Freunde Berl. Mag. 2: 320 (1808).
Arundinaria glaucescens (Willd.) Beauv., Ess. Agrost. 144, 152 (1812).
Bambusa nana Roxb., Fl. Ind. 2nd ed., 2: 199 (1832).
Bambusa glaucescens (Willd.) Sieb. ex Munro in Trans. Linn. Soc. 26: 89 (1868); Holttum in Kew Bull. 11 (2): 207-211 (1956), Gard. Bull. Sing. 16: 67 (1958).

Densely tufted bamboo; culms erect or arching, 2.5-7 m tall, diameter 1.5-2.5 cm at the base, with relatively thick walls, internodes 30-60 cm long, covered with white wax when young, smooth, glabrous, nodes not swollen. Branch complement several at each node. Culm sheaths smooth, glabrous, light green when young, becoming stramineous, apex rounded; blades narrowly to broadly triangular, tapering to the tip, widened at the base to the full width of the sheath apex, extending to narrow auricles; ligule very short. Leaf blades 6.5-14 x 1-1.5 cm, base cuneate or rounded, glabrous; sheaths glabrous; auricles

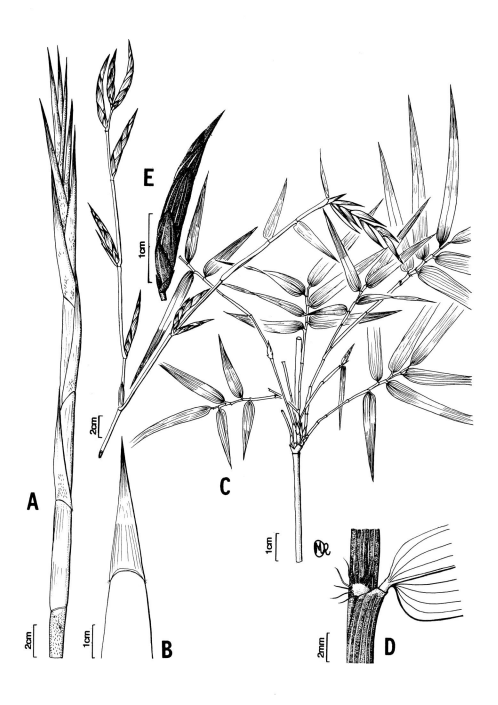

Fig. 3. *Bambusa multiplex*. A, Culm shoot. B, Culm sheath, inner (adaxial) side. C, Leafy branches. D, Leaf sheath, detail. E, Flowering branches and pseudospikelet. (A-D from fresh material; E from SAN 124360.)

small with fine bristles. Inflorescence usually borne on leafless branches. Spikelets slender, more or less cylindrical when young, 15-22 mm long (or more), consisting of 2 glumes and up to 10 florets; lemma glabrous; palea glabrous, keels slightly fringed; lodicules 3.

Distribution. Native to China, introduced everywhere, especially in the tropics as a popular hedge plant.

Habitat. This bamboo can be found growing on all kinds of soil, from low elevation to about 2000 m altitude in the tropics. In China this bamboo will stand low temperature planted outdoors in Hangzhou.

Vernacular names. *Bambu China*, *buluh pagar* (Malay).

Uses. *B. multiplex* is planted as hedges all over the tropics. The culms are often used as fishing rods.

Notes. There are several varieties of *B. multiplex*. One of them is the common hedge bamboo in Sabah.

4. Bambusa tuldoides Munro

(Latin, resembling *Bambusa tulda*)

Munro in Trans. Linn. Soc. 26 : 70 (1868).

Synonym :
Bambusa ventricosa McClure in Lingnan Sci. Journ. 17: 57 (1938); Holttum in Gard. Bull. Sing. 16: 70 (1958).

Open-tufted bamboo; culms about 2 m (normally) or 50 cm (in dwarf state) long, diameter 2-2.5 cm at the base, dull or glossy green, internodes cylindrical or swollen; branch complement of few branches, with the middle branch dominant. Culm sheaths glabrous, green becoming stramineous; blades erect, triangular, slightly narrowed at the base, usually glabrous; auricles small but prominent, with bristles along the edge. Leaf blades 10-15 x 2-2.5 cm, base rounded, lower surface soft hairy; sheaths usually glabrous; auricles small with long spreading bristles along the edge.

Distribution. Native to southern China, introduced elsewhere as an ornamental plant, especially the dwarf variety.

Notes. In Sabah this species is not common. A plant of the dwarf variety will retain the dwarf character if planted in a pot, but the clump will produce cylindrical culms when planted in the ground.

5. **Bambusa vulgaris** Schrader ex Wendland   (Fig. 4)

(Latin, common)

Wendland, Collect. Pl. 2: 26 (1810); Gamble in Ann. Roy. Bot. Gard. Calc. 7: 43 (1896); Holttum in Gard. Bull. Sing. 16: 63 (1958); Soderstrom & Ellis, Smiths. Contrib. Bot. 72: 39 (1988).

Open tufted bamboo; culms 10-20 m tall, diameter 4-10 cm at the base, with relatively thick walls, dark glossy green or yellow with green stripes, slightly zig-zag, or the nodes not horizontal but slightly angled, prominent, lowermost ones often bearing aerial roots, internodes 20-45 cm long, smooth and shiny when mature; young shoots light or dark green, covered with black or dark brown hairs. Branch complement of few branches with the middle branch dominant, branches borne from lower nodes upwards or from midculm nodes upwards, arranged alternately along the culm, the dominant middle branches prominent in young culm, together forming a gigantic fan-like structure. Culm sheaths light to glossy green in the green variety or yellow in the yellow variety, becoming stramineous, covered with dark brown or black hairs; blades broadly triangular, with pointed apex, erect, adaxial surface covered with dark brown hairs especially near the base, margins with short curly bristles especially near the base; auricles large, with long curly bristles along the edge; ligule short, irregularly toothed. Leaf blades 9-30 x 1-4 cm, glabrous; sheaths glabrous; ligule very short; auricles not prominent. Inflorescence comprising clustered spikelets borne on leafless branches. Spikelets 20-35 mm long, consisting of 5-10 florets, rachilla internodes short, glabrous; glumes 1 or 2, glabrous; lemmas short-hairy near the apex, otherwise glabrous; paleas 2-keeled, keels hairy.

Distribution. Found all over the tropics, of unknown origin, wild, naturalized or planted. Common in Sabah.

Habitat. *Bambusa vulgaris* will grow on any kind of soil, from low elevation to 300 m altitude, but usually found commonly along rivers.

Vernacular names. *Tamalang* or *tambalang* (Dusun, Murut, Kadazan). The yellow-culm variety is called *tamalang silau* by the Dusun people and *bambu kuning* in Malay.

Uses. The culms are strong but not straight, and are used for poles, building bridges, masts, rudder, etc.

Notes. There are several varieties of *B. vulgaris*. The most common varieties in Sabah are the green-culm and the yellow-culm varieties. The yellow variety is usually planted as an ornamental, whereas the green variety is found growing spontaneously or naturalised by roadsides or along rivers and paddy fields. The other variety is *B. vulgaris* cv. *Wamin* McClure, a variety or a cultivar with short inflated internodes, found planted in Sabah as an ornamental. The origin of this cultivar is probably China.

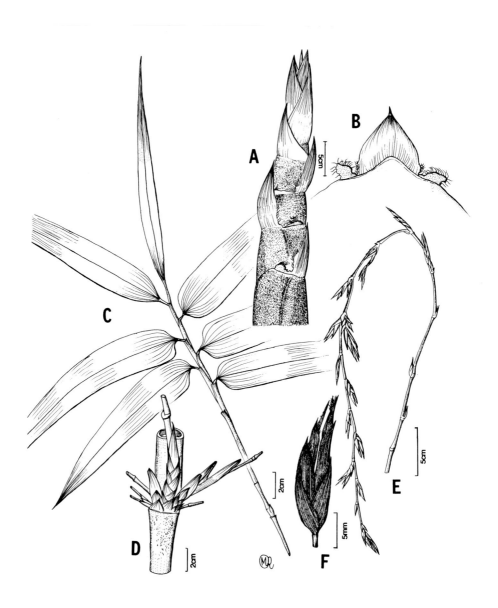

Fig. 4. *Bambusa vulgaris*. A, Culm shoot. B, Culm sheath, inner (adaxial) side of upper part. C, Leafy branch. D, Branch complement. E, Flowering branch. F, Spikelet. (All from fresh material except E, F, from SAN 41195.)

The species can be recognised from a distance by the fan-like structure of the young culms with their branches. In general it is almost impossible to identify a bamboo by its culm, but the culms of *B. vulgaris* can be easily recognised, because they are not straight and the nodes are not horizontal.

B. vulgaris is easily propagated; a piece of culm cutting or branch cutting will produce a new propagule or plant under favourable conditions without much care. The culms are often used for *dayung* (boating poles), and because this bamboo is easily available the *dayung* would be thrown away after use. Such pieces of culm would survive and produce roots and new growth then becomes established on river banks. This must have happened everywhere over centuries. This species is one of the faster-growing bamboos. The *rebung* is good to eat, but rarely sold in the market, because it develops rapidly overnight and becomes too old or fibrous to use as a vegetable.

6. **Bambusa** sp.

During a survey of Sabah bamboos in 1979, some clumps of this erect bamboo were seen cultivated in a village garden near Kudat. There, the culms, two joined together, were used as poles for collecting or picking coconuts from a tall palm. The culms are about 7 m tall, and about 3 cm in diameter, slender, and the branches are borne from the midculm nodes upwards. This bamboo looks like *Bambusa mutabilis* McClure from South China, but the internodes are longer than those in *B. mutabilis*. Herbarium specimens were not collected during this survey. There should be further investigation of this bamboo.

DENDROCALAMUS Nees

(Greek: *dendron*, tree; *kalamos*, reed; tree-like reeds).

Nees in Linnaea 9 : 476 (1834); Holttum in Gard. Bull. Sing. 16 : 86 (1958).

Densely tufted bamboos of medium to large size, culms erect, of various diameters, with drooping tips, straight, usually hollow with small lumen, with moderately thick walls. Branches usually borne from midculm nodes, rarely from the lower part (*D. strictus*). Culm sheath usually covered with pale hairs, and with waxy powder; blades rigid; auricles present or absent. Inflorescence borne on branches of leafless culms. Pseudospikelets in groups at nodes of flowering branches; spikelet comprising 1-6 perfect florets, and an apical sterile or reduced floret borne on an elongating extension of the rachilla.

Distribution. The genus is mainly found in tropical Asia, with its centre of distribution in mainland Asia (i.e. from the Indian subcontinent to Vietnam). So far there is only one species recorded in Sabah, namely *D. asper*.

Dendrocalamus asper (Schultes f.) Backer ex Heyne

(Latin: *asper*, rough, uneven)

Heyne, Nutt. Pl. Ned. Ind. ed. 2, 1: 301 (1927); Holttum in Gard. Bull. Sing. 16: 100 (1958); Dransfield & Widjaja in PROSEA, a selection, 109-112 (1989).

Synonyms:
Bambusa aspera Schultes F., Syst. Nat. 7: 1352 (1830).
Dendrocalamus flagellifer Munro in Trans. Linn. Soc. 26: 150 (1866).
Gigantochloa aspera (Schultes f.) Kurz in Ind. For. 1: 340-341 (1876).

Densely tufted bamboo; culms erect with pendulous tips, 20-30 m tall, diameter 8-20 cm, when young densely covered with velvety fine golden brown hairs, lower internodes 10-20 cm long, gradually increasing and upper ones 30-50 cm long, walls 11-20 mm thick, with white wax below nodes, lowermost nodes swollen bearing dense aerial roots, upper nodes not so prominently swollen; young shoots covered with dark brown to black hairs. Branch complement of few branches and borne from the midculm nodes upwards, primary branch dominant. Culm sheaths covered with dark brown hairs, those of young shoots and of the lower part of the culm about 20 cm long, 20 cm wide at the base with small deflexed blades, those of the upper part of the culm up to 40 cm long, 25 cm wide at the base, with light brown hairs, blades erect first then deflexed, 25 x 3 cm; auricles prominent, bearing bristles along the edge; ligule 10 mm tall, lacerate. Leaf blades 30 x 2.5 cm, petiole 4 mm long, glabrous above, hairy below, base shortly attenuate; sheaths glabrous or with scattered appressed pale hairs; auricles not present; ligule very short. Inflorescences borne on leafless branches of leafy or leafless culms, axes glabrous or hairy. Spikelets 6-9 mm long, about 5 mm wide, consisiting of 1-2 glumes, 4-5 florets; lemmas fringed with pale hairs towards the apex and along the margins, obtuse; paleas with fringe along the keels.

Distribution. The origin of *D. asper* is not certain, but it is somewhere in South East Asia, where it is commonly planted.

Habitat. This species is commonly found at 400-500 m, although it grows well from low altitude to 1500 m.

Vernacular name. *Buluh betung* (Malay, Indonesian).

Uses. The culms are very strong and durable, and are used for building material. The *rebung* is very good to eat.

Notes. During a survey of Sabah bamboos in 1979, this species was not observed. This species was reportedly found growing on hills in Long Pasia (Tony Lamb, pers. comm.). It is easily propagated from branch cuttings.

DINOCHLOA Büse

(Greek: *deinos*, fearful, or marvelous; *chloa*, grass).

Büse in Miquel, Plantae Jungh. 388 (1854); Holttum in Gard. Bull. Sing. 16: 31 (1958); Dransfield in Kew Bull. 36(3): 617 (1981).

Open-tufted climbing bamboos; culms zig-zag, usually solid, rarely hollow with small lumen, smooth or rough when young, usually purplish, rarely green, covered with white wax, becoming smooth and green by age. Young shoots purplish or green, covered with white wax. Culm sheaths with a distinctive basal portion which is usually hard, rugose and coarse, glabrous or hairy, the main part of the sheath smooth, glabrous or hairy; auricles present or absent; ligule very short or long and laciniate; blades ovate or ovate-lanceolate, glabrous or hairy near the base. Branches 3-18 at each node with the middle branch dormant (but developing in a way similar to the main culm bearing it, when the apex of the main culm is damaged). Leaf blades very large (30-35 cm long) or small (up to 7 cm long), smooth or rough, glabrous or hairy; auricles present or absent. Inflorescence iterautant in a form of a huge leafless flowering branch of 2-3 m long, or a short flowering portion terminating a leafy branch. Spikelets 2-9 mm long, comprising 2-3 glumes and one floret; stamens 6, anthers with apiculate tips; ovary glabrous, stigmas 3. Fruit globose, subglobose or obclavate, up to 7 mm (but c. 30 mm in one species) in diameter, smooth or rugose; pericarp thick and fleshy; endosperm small or much reduced; embryo with large scutellum filling up the fruit cavity and containing starch grains; plumule and radicle lateral.

Distribution. The genus *Dinochloa* is confined to South East Asia, that is from the Andaman Islands and southern Thailand, to Malaysia, Indonesia and the Philippines. There are about 20 species, only 14 are named; of the nine species known in Sabah, four are not known elsewhere.

Habitat. All species are found growing wild, scattered in the lowland and hill dipterocarp forests (altitude 1200 m), but they become weeds in logged-over or burnt forest, secondary forest or forest margins, and there is much concern that they may prevent regeneration of commercial timber species.

Local name. *Dinochloa* species are frequently called *wadan* by local people.

Notes. *Dinochloa* has several interesting features. The climbing habit is a most unusual feature in bamboos, known only in three other bamboo genera, namely *Melocalamus*, *Chusquea* and *Nastus* of Madagascar. In *Dinochloa* the young shoots or young culms are first erect and straight, but when they develop and elongate the lower parts scramble on the forest floor among other plants and the upper parts climb or twine around tree trunks. Although the climbing habit is probably governed internally, the zig-zag internodes help the culm twine around tree trunks and the rough bases of the culm sheaths help the culm to cling onto tree bark.

In all bamboo species the apex of the young culm is soft and is easily damaged, and in most of them the culm will die sooner or later. In *Dinochloa* if the apex of the culm is damaged, the dormant primary branch axis of each node will grow and elongate to replace the main culm. If the apex of this new culm is damaged, its dormant buds will also grow and elongate, and so on. In this way the culms will survive in the forest. It is very difficult though to measure or to know the length of individual culms.

Because the fruit has almost no endosperm, and the food is stored in its large scutellum, the fruit has no resting period and germinates as soon as it reaches maturity. It sometimes even germinates while still on the parent plant (*i.e.*, it is viviparous), as observed in *D. prunifera* in 1979. The plumule and radicle emerge from the lateral side of the fruit by breaking through the pericarp during germination. When a clump or a plant of a *Dinochloa* species produces flowers, it always produces large numbers of fruits. Seedlings can often be seen scattered on the forest floor under the plant.

As with other species of bamboo, once a clump of *Dinochloa* is established, it is very difficult to eradicate. Local people do not use the culms of *Dinochloa* species, but when there is no other bamboo available, the culms are used for making rough baskets for carrying stones from the river. Dusun people sometimes use the watery sap from freshly cut solid culm portions as an eye drop. The leaves of *D. trichogona* are sometimes used (see species notes).

Key to the species

1. Leaf blades large, 25-45 x 5-7 cm, glabrous, smooth; culm usually solid 2
    Leaf blades small to medium size, up to 20 cm long, no more than 3.5 cm wide, hairy or with prickle hairs, rough; culms solid or hollow 3

2. Base of culm sheath covered with golden brown hairs, auricle with long curly bristles, ligule 15 mm long, deeply laciniate, blades hairy adaxially especially near the base; leaf blades with auricles and long ligule (11 mm long); inflorescence axes hairy; fruit smooth; widespread .... *D. trichogona*

Base of culm sheath glabrous, auricles without bristles, ligule short, 2 mm long, entire, blades glabrous; leaf blades without auricles, ligule short; inflorescence axes glabrous; fruits rugose; so far found only around Poring .... *D. sublaevigata*

3.   Leaf blades 15-20 x 2-3 cm 4
        Leaf blades small and narrow, 13-15 x 1.8 cm 6

4.   Young shoots purplish, blades of culm sheaths erect or deflexed; fruits globose, diameter 5-9 mm, smooth 5
        Young shoots green covered with thick white wax, blades of culm sheaths erect first then deflexed; fruits subglobose, diameter 25 mm, rugose (so far found only in Telupid, on ultramafic soil) ...... *D. prunifera*

5.   Culm sheath auricles not present; leaf blades scabrid on one side of the lower surface near the margin, otherwise glabrous, sheath glabrous (auricles not seen) (widespread) .... *D. scabrida*
        Culm sheath auricles present, with bristles; leaf blades hairy below, sheaths densely hairy (auricles present with long bristles) (very common on east coast only) .... *D. darvelana*

6.   Slender bamboos; culms 7-8 mm in diameter; leaf blades hairy 7
        Robust bamboo; culm 15-20 mm in diameter, always hollow; culm sheath with appressed white hairs, glabrous later, base with stiff light brown hairs, blades deflexed; leaf blades glabrous (found in Kudat, P. Banggi and Palawan in the Philippines) .... *D. robusta*

7.   Blades of culm sheaths usually erect, ligule very short, entire 8
        Blades of culm sheaths usually deflexed, ligule 10 mm long, deeply laciniate; leaf blades with prominent auricles with long bristles; inflorescence axes hairy (common around Telupid)............ *Dinochloa sp.*

8.   Culms usually hollow with thin walls; blades of culm sheaths narrow; inflorescence axes glabrous; fruit obclavate, light green or creamy white when fresh, with purplish stripes (near Telupid, on ultramafic soil) .... *D. obclavata*
        Culms usually solid; blades of culm sheaths ovate-lanceolate, widened at the base; inflorescence axes usually hairy; fruit globose, light green without purplish stripes (from Sipitang to Sarawak) .... *D. sipitangensis*

1. Dinochloa darvelana S. Dransf. (Fig. 5)

(named after Darvel Bay, around where this species is common)

Dransfield in Kew Bull. 44(3) : 435-437 (1989).

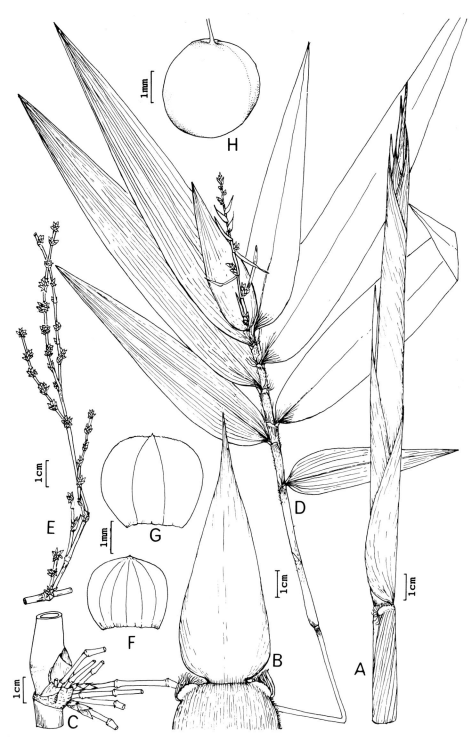

Fig. 5. *Dinochloa darvelana*. A, Young shoot. B, Culm sheath with blade and auricles. C, Branch complement. D, Leafy branch. E, Part of flowering branch. F, Lemma. G, Palea. H, Fruit. (A-B from SD 836; C-H from SD 834.)

Culm usually hollow or with small lumen, smooth, internodes 35-40 cm long, 1-1.5 cm in diameter; young shoots purplish, with thin white wax. Culm sheath purplish, glabrous, 10-13 cm long, one margin usually hairy, base glabrous; auricles about 6 mm long, with long bristles; blades ovate, tapering, erect, glabrous, 10-14 cm long, 4 cm wide at the base; ligule very short, entire or minutely and irregularly toothed. Leaf blades 15-23 X 1.5-3 cm, glabrous above, glabrous or minutely pubescent below, sheaths hairy, auricle prominent but easily shed, with long bristles. Inflorescence about 2 m long, shorter ones of 10-15 cm long terminating leafy branches, axes glabrous or pubescent. Spikelet about 3 mm long, glabrous. Fruits globose, smooth, green, sometimes tinged with red or purple, up to 9 mm in diameter.

Distribution. Sabah: Darvel Bay, Bukit Silam, Pulau Sakar. Indonesia: East Kalimantan (Samarinda).

Habitat. Forest or forest margins on ultramafic rock, and primary forest on volcanic rock.

Notes. *D. darvelana* is closely related to *D. scabrida*, a widespread species in Borneo. It differs, however, from *D. scabrida* by its large auricles of the culm sheath, the glabrous (or pubescent), smooth leaf blades and the sheath bearing auricles with long bristles.

2. **Dinochloa obclavata** S. Dransf. (Fig. 6)

(Latin, club-shaped but attached by thicker end, alluding to its fruit shape)

Dransfield in Kew Bull. 36(3): 620 (1981).

Culms about 7 mm in diameter, thin-walled, glabrous, smooth, internodes 20-30 cm long; young shoot purplish with white wax; secondary branches up to 18 at each node. Culm sheaths glabrous, purplish when young, covered with white wax, narrowing distally, up to 21 cm long; auricles not present; ligule 1 mm long; blades narrow, glabrous, usually erect, somewhat deflexed by age. Leaf blades rough, hairy when young, becoming glabrescent, 13-20 x 1.2-1.8 cm; sheath with appressed stiff hairs when young, becoming glabrous, margins hairy; auricles 1 mm long, with long bristles, easily shed; ligule very short with long bristles. Inflorescence about 1.5 m long, axes glabrous. Spikelets 2-3.5 mm long, flattened, ovate-lanceolate. Fruits 1.5 cm long, 9 mm diameter, obclavate, smooth, with pointed apex, light green or creamy white when fresh, with purple stripes.

Distribution. So far this species is known only from around Telupid.

Habitat. Forest on ultramafic soil, at *c.* 50 m altitude.

Notes. *D. obclavata* can be distinguished from other species of *Dinochloa*

26

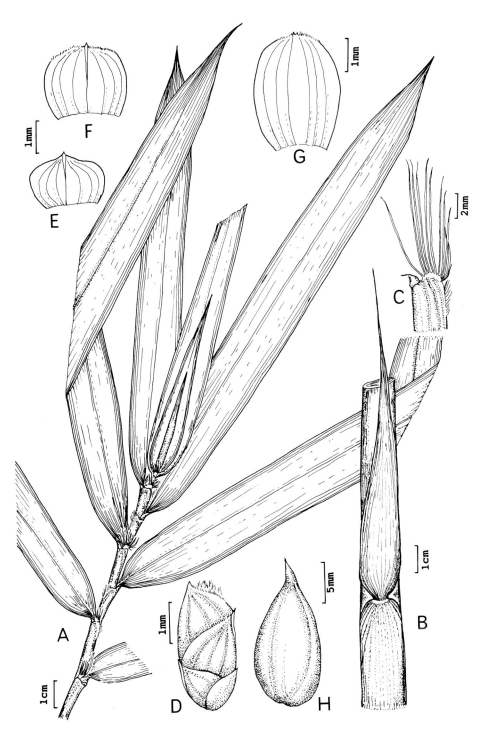

Fig. 6. *Dinochloa obclavata*. A, Leafy branch. B, Culm sheath. C, Auricle of leaf blade. D, Spikelet. E, Upper glume. F, Lemma. G, Palea. H, Fruit. (A, C-G from SD 784; B from SD 777.)

27

occurring in Sabah in having small hairy leaf blades, many small secondary branches at each node, and by its small, flattened, ovate-lanceolate spikelets.

3. **Dinochloa prunifera** S. Dransf.   (Fig. 7)

(Latin, bearing plums, in reference to the plum-shaped fruit)

Dransfield in Kew Bull. 36(3): 622 (1981).

Culm about 1 cm in diameter, hollow with thick walls, smooth, internodes about 25 cm long; young shoots green covered with thick white wax. Culm sheaths green, glabrous, about 15 cm long, covered with white wax, hairy along the margins; auricles prominent, 1-1.5 mm long, with spreading long white bristles (bristles up to 12 mm long); blades green, ovate to ovate-lanceolate, more or less cordate at the base, 13 x 2.5 cm, erect or deflexed, glabrous, covered with white wax when young; ligule very short. Leaf blades 17-20 x 2-3.2 cm, rounded at the base, glabrous, rather rough; auricles small, with long bristles (bristles up to 8 mm long); ligule very short. Inflorescence about 3 m long, axes pubescent. Spikelets 7-9 mm long, elongate, flattened, oblique. Fruits subglobose, 17-30 x 25 mm, rugose, green when fresh.

Distribution. So far found only around Telupid.

Habitat. Forest on ultramafic soil, especially where disturbed, such as along logging roads.

Notes. *D. prunifera* has the largest fruit known in the genus. In the field this species can be recognised by its green waxy young shoots and long bristles on the auricles of the culm sheaths. *D. prunifera* was first collected in 1979, when it was found flowering and fruiting in a forest margin about 75 km from Sandakan on the way to Telupid. It was not common then. In 1986 it has become rarer, and only one clump was seen climbing among the secondary growth by the roadside.

4. **Dinochloa robusta** S. Dransf.   (Fig. 8)

(Latin, referring to the robust habit)

Dransfield in Kew Bull. 47(3): 402 (1992).

Extremely robust climbing bamboo. Culm hollow, 15-20 mm diameter, with about 5 mm thick walls, internodes rather rough with scattered stiff appressed hairs, about 30 cm long; young shoots purplish green. Culm sheaths usually glabrous, but occasionally with appressed white hairs, base covered with stiff light brown hairs; ligule very short, entire; auricles not present; blades deflexed, broadly ovate, tapering to the tip, base cordate, 11-18 x 5.5 cm, adaxially pubescent, especially near the base, glabrous abaxially. Leaf blades 9.5-17 x 1.5-2.5 cm, smooth, glabrous; ligule very short, usually without bristles.

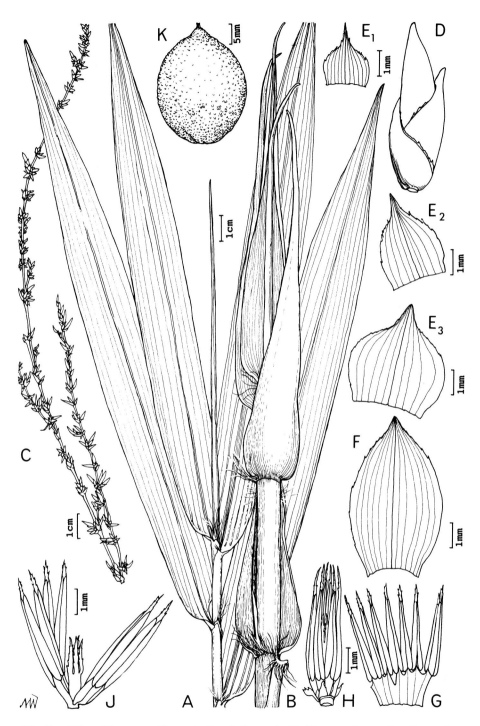

Fig. 7. *Dinochloa prunifera.* A, Leafy branch. B, Young shoot showing culm sheath. C, Part of flowering branch. D, Spikelet (lowermost glume removed). E1-E3, Glumes. F, Lemma. G, Stamens with fused filaments (in abnormal condition). H-J, Stamens and ovary. K, Fruit. (All from SD 776.)

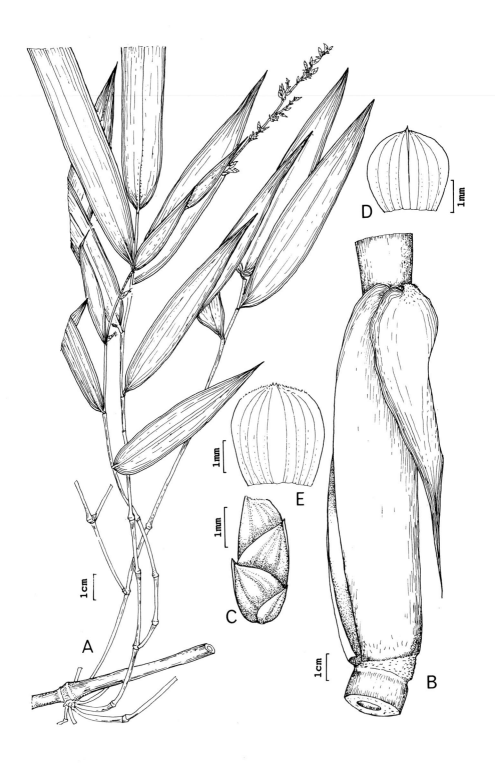

Fig. 8. *Dinochloa robusta*. A, Leafy branch. B, Culm sheath. C, Spikelet. D, Lemma. E, Palea. (A, C-D from Balajadia 3757; B from SD 752.)

Inflorescence about 2 m long, or 8-10 cm long terminating leafy branches, axes glabrous. Spikelets 2.5-3 mm long. Fruit not seen.

Distribution. Sabah: Kudat, P. Banggi. Philippines: Palawan.

Habitat. Edge of forest or secondary forest on ultramafic soil, or alluvial soil.

Notes. This species resembles *D. sipitangensis* in having small leaf blades, and *D. sublaevigata* in its broadly ovate blades of the culm sheaths.

5. **Dinochloa scabrida** S. Dransf.   (Fig. 9)

(Latin, scabrid, alluding to the rough undersurface of the leaf blades)

Dransfield in Kew Bull. 36(3): 628 (1981).

Culms solid, or rarely with small lumen, 2 cm diameter, internodes 20-25 cm long, smooth; young shoots purplish. Culm sheaths glabrous, purplish, covered with white wax; blades usually erect, 8.5-10 x 2.5 cm, narrowly triangular, cordate, tapering, glabrous; ligule 2 mm long, entire; auricles not present. Leaf blades 12-25 x 1.5-2.5 cm, lanceolate, scabrid on one side of the lower surface near the margin, otherwise glabrous, base attenuate; sheath glabrous. Inflorescence 3 m long, or 15 cm long terminating leafy branches, axes glabrous. Spikelets 2.5 mm long. Fruits globose, smooth, up to 5 mm diameter.

Distribution. Sabah: widespread; Sarawak. Indonesia: East Kalimantan.

Habitat. Lowland dipterocarp forests, along roads or forest margins.

Notes. *D. scabrida* is probably the commonest species of *Dinochloa* in Borneo, and has become a weed in burnt and logged-over forests.

6. **Dinochloa sipitangensis** S. Dransf.   (Fig. 10)

(named after Sipitang, the type locality)

Dransfield in Kew Bull. 36(3): 620 (1981).

Culms solid or with a small lumen, 9 mm diameter, with appressed white hairs when young, becoming smooth, internodes about 25 cm long; young shoots purplish. Culm sheaths smooth, glabrous, margins hairy; ligule very short; blades usually erect, narrow, 3-7 x 0.5 cm, widened at the base, usually glabrous; auricles not seen. Leaf blades 14-15.5 x 1.3-1.5 cm, hairy on both surfaces, especially when young; ligule short, laciniate; sheaths hairy on the back. Inflorescence about 1.5 m long, axes usually hairy. Spikelets up to 3 mm long. Fruits 7-8 mm diameter, globose, smooth.

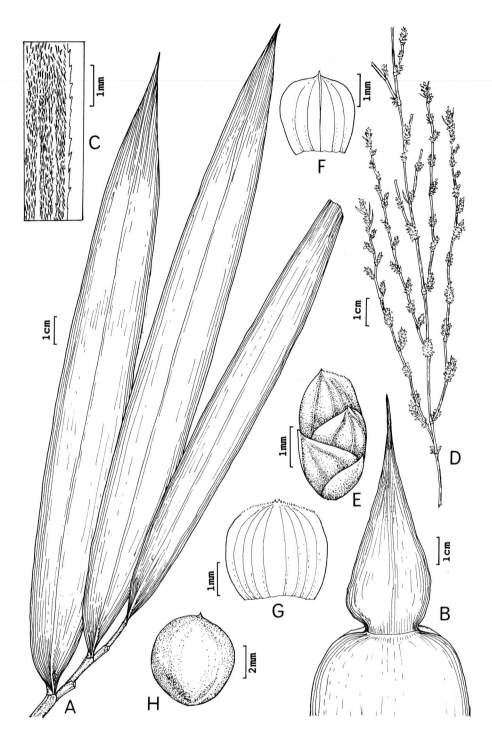

Fig. 9. *Dinochloa scabrida*. A, Leafy branch. B, Culm sheath. C, Lower part of leaf blade. D, Part of flowering branch. E, Spikelet. F, Lemma. G, Palea. H, Fruit. (All from SD 746.)

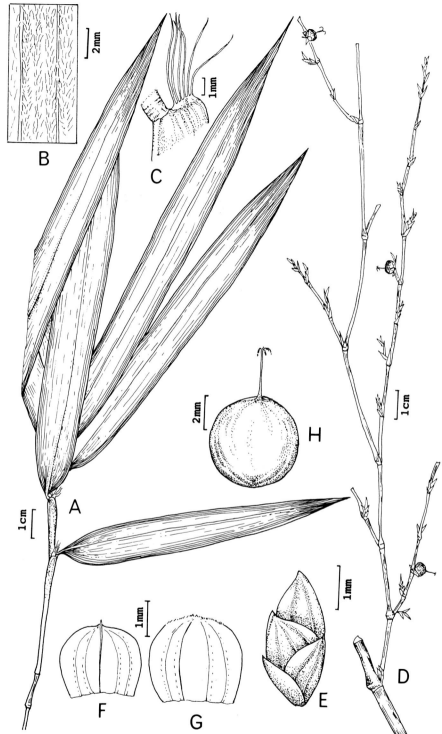

Fig. 10. *Dinochloa sipitangensis*. A, Leafy branch. B, Lower surface of leaf blade. C, Auricle of leaf blade. D, Part of flowering branch. E, Spikelet. F, Lemma. G, Palea. H, Fruit. (All from SAN 43276.)

Distribution. Sabah (especially in west coast near Sipitang); Sarawak. Brunei.

Habitat. In forest on sandy soil, or forest margins by roadsides.

7. **Dinochloa sublaevigata** S. Dransf. (Fig. 11)

(Latin, alluding to the culm and culm sheaths, which are not completely smooth or glabrous)

Dransfield in Kew Bull. 36(3): 626 (1981).

Culms usually solid, 2 cm diameter, with stiff hairs when young, becoming smooth and glabrous by age, internodes 30-35 cm long; young shoots and young branches purplish, glabrous. Culm sheaths purplish with short appressed hairs when young, becoming glabrous, margins with long curly hairs especially when young, base glabrous; auricles 3 x 8 mm, easily shed, without bristles; ligule 2 mm long, entire; blades 8.5 x 2-2.5 cm, broadly ovate, tapering to a fine tip, glabrous, usually deflexed. Leaf blades 30-40 x 4.5-6.5 cm, smooth, base slightly attenuate, cross veins prominent below; ligule very short; auricles not present; sheaths glabrous, margins with short white hairs. Inflorescence 2-3 m long, axes hairy or somewhat glabrous. Spikelets 4 mm long. Fruits up to 9 mm long, diameter 8-9 mm, globose, rugose.

Distribution. Sabah : northern and eastern parts.

Habitat. Hill dipterocarp forest, forest margins along roadsides.

Notes. *D. sublaevigata* is commonly seen in the area around Poring and Bukit Ampuan. Along the road between Ranau and Poring one can see and recognise this species with broad leaf blades, climbing on trees.

8. **Dinochloa trichogona** S. Dransf. (Fig. 12)

(Greek, alluding to the hairy node or base of the culm sheath)

Dransfield in Kew Bull. 36(3): 624 (1981).

Culms solid, diameter 2-3 cm, rough with scattered appressed hairs when young, becoming smooth, internodes 20-25 cm long; young shoots purplish, hairy. Culm sheaths purplish with brown hairs when young, brownish and glabrous by age, margins hairy, base covered with dense, erect brown hairs; auricles 5-15 mm long, with long curly bristles (bristles 7 mm long); blades erect first, then deflexed, 11.5 x 2.5 cm, cordate, long acuminate or tapering to fine long tip, glabrous and smooth abaxially, densely hairy adaxially, especially near the base; ligule about 15 mm tall, deeply laciniate. Leaf blades 25(-45) x 7 cm, with prominent cross veins below, smooth, glabrous, base rounded; auricles 5 x 7 mm, with long curly bristles (bristles 10 mm long); ligule about 11 mm tall,

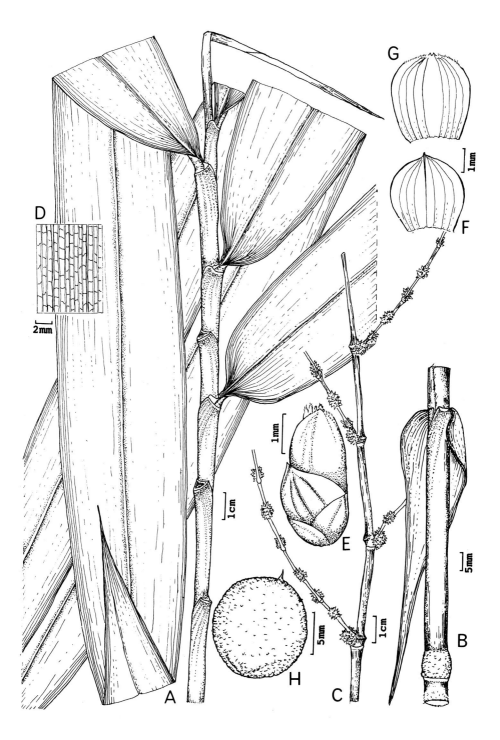

Fig. 11. *Dinochloa sublaevigata*. A, Leafy branch. B, Culm sheath. C, Part of flowering branch. D, Lower surface of leaf blade. E, Spikelet. F, Lemma. G, Palea. H, Fruit. (All from SD 720.)

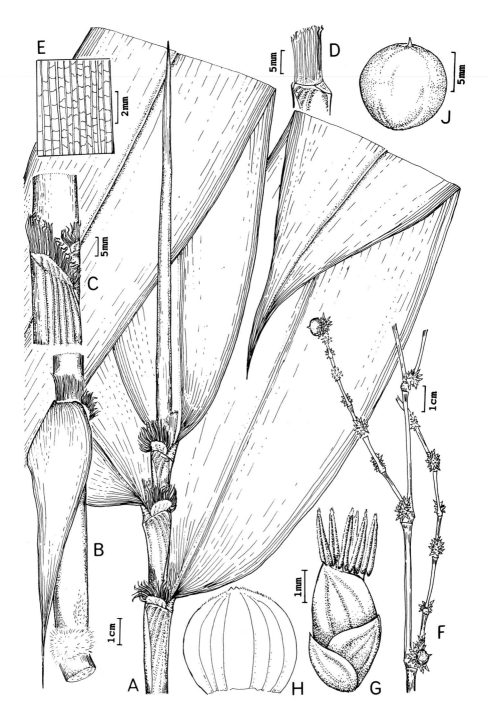

Fig. 12. *Dinochloa trichogona*. A, Leafy branch. B, Culm sheath. C, Auricle of culm sheath. D, Ligule of leaf blade. E, Lower surface of leaf blade. F, Part of flowering branch. G, Spikelet (lower glume removed). H, Palea. J, Fruit. (All from SD 747.)

laciniate, easily shed; sheaths hairy along the margins. Inflorescence 2-3 m long, axes usually pubescent. Spikelets 2-3 mm long. Fruits more or less globose, smooth, green when young, 10 mm long, 9-10 mm diameter.

Distribution. Sabah; Sarawak. Brunei.

Habitat. Lowland and hill dipterocarp forest, forest margins along roadsides, often becoming weeds in logged forests.

Notes. *D. trichogona* can be easily recognised by its broad, smooth leaf blades and hairy base of the culm sheath. Occasionally, however, the base of the culm sheath is not markedly hairy, but the species still can be recognised from its long laciniate ligule of the culm sheath and of the leaf blade.

Black ants are often seen foraging around the margins of blades and auricles of the culm sheaths on young shoots or branches. It is not known whether there is any significance in this association; it is possible that these ants help to protect the young shoots which are fragile. Stingless bees are also often seen visiting the flowers and collecting pollens.

D. trichogona and *D. sublaevigata* are closely related species. The leaf blades of both species are used by local Chinese for wrapping Hokkien *Bak Chang* (made of glutinous rice), as the blades are large and smooth (no prickle hairs). These two species are often found growing together in the area around Ranau. *D. trichogona* differs from *D. sublaevigata* by having hairy young shoots, prominent auricles on both the culm sheath and the leaf sheath, a hairy base of the culm sheath, and a smooth fruit. *D. trichogona* is also found growing together with *D. scabrida* in the area between Sandakan and Telupid. Here, occasionally, plants which are intermediate between these two species, can be found. There is no record, however, that different *Dinochloa* species can hybridise.

9. **Dinochloa** sp. (as yet unnamed)

This species is found abundantly near Telupid after the forest was destroyed by the fire in 1983-1986. Fruits have not been found. The small leaf blades resemble those of *D. sipitangensis* or *D. obclavata*. At present, this taxon is kept separate from these two species, until the plant is found producing flowers and fruits.

GIGANTOCHLOA Kurz ex Munro

(Greek : *gigas*, large; *chloa*, grass)

Munro in Trans. Linn. Soc. 26: 123 (1868); Backer, Handb. Fl. Java 2: 273 (1924); Holttum in Gard. Bull. Sing. 16: 104 (1958); Widjaja in Reinwardtia 10(3): 301 (1987).

Densely tufted bamboos of medium to large size. Culms erect, straight, usually hollow with moderately thick walls, nodes usually not swollen, young culms often covered with hairs or rarely with white wax, glabrous and smooth at maturity. Branch complement of few branches, borne from midculm nodes upwards, the middle branches larger than the others. Culm sheaths usually covered with pale brown, rarely white, to dark brown hairs; blades broadly lanceolate to ovate-lanceolate, usually deflexed or spreading, glabrous or hairy; auricles small to large, with or without bristles. Leaf blades often with short petiole. Inflorescences borne on branches of leafless culms. Pseudospikelets in groups at nodes of flowering branches, usually hairy; spikelets consisting of 2-5 glumes, 2-5 florets, terminated by an empty lemma, rachilla internodes very short.

Distribution. The genus is mainly found in tropical Asia, and consists of about 21 species. So far two species occur in Sabah; they are *G. levis* and *G. balui*. It is possible that there are other species not recorded here planted in villages.

Notes. Most species of *Gigantochloa* are useful to local people and are planted in villages for everyday use. For this reason, therefore, many of the species are found only in cultivation, and some grow spontaneously or are naturalised around villages, in forest margins, and on wastelands. They are almost absent in primary forest, but they can form a pure bamboo forest in an area where the forest has been completely destroyed or heavily logged.

Key to the species

Usually large bamboo; young shoot purplish or brownish green, covered with dark brown hairs; culm sheaths green to purplish green, often tinged with orange, covered with dark brown hairs; blades erect first then spreading, hairy above; auricles large, with bristles along the edge; ligule laciniate, with bristles along the edge .... *G. levis*

Usually medium-size bamboo; young shoot pale green, with white hairs; culm sheaths pale green, covered with white hairs; blades erect then spreading, glabrous; auricles small, without bristles; ligule short, entire .... *G. balui*

1. **Gigantochloa balui** K. M. Wong (Fig. 13)

(after the local name *balui*)

Wong, For. Dept. Occ. Papers, Brunei 1: 1-10 (1990).

Medium-size bamboo; culms 7-12 m tall, diameter 5-8 cm, with moderately thin walls, internodes 30-40 cm long, smooth, light green to glossy green, often with white wax when young, nodes not swollen; young shoots with white hairs. Culm sheaths 17-20 cm long, 14-19 cm wide at the base, 4-5 cm wide at the apex, covered with appressed fine white hairs, becoming glabrous; blades 12-19 cm long, 2.5 cm wide near the base, tapering, pubescent abaxially, first erect then deflexed or spreading; auricles low, but forming a long rim of about 2 cm long from the base of the blade to the margin of the sheath apex; ligule 2-3 mm tall, fringed along the edge. Branch complement of few branches, developing from midculm nodes upwards. Leaf blades usually glabrous, 23-35 x 2-4 cm, base attenuate, petiole 5-10 mm long, usually soft- hairy, especially on the lower surface; sheaths glabrous; auricles not prominent; ligule very short. Inflorescences borne on branches of leafy or leafless culms, axes glabrescent. Spikelets laterally flattened, 10-12 mm long, consisting of 3-5 glumes, 2-3 florets, and one empty lemma; glumes and lemmas glabrous on the back, upper part of the margins fringed with light brown hairs, with pointed, out-curved tip; paleas hairy along the keels.

Distribution. Sabah; Sarawak. Brunei. Found planted in villages and growing spontaneously in wastelands along roadsides or around villages.

Habitat. River banks or wastelands by roadsides.

Vernacular names. *Balui* (Bajau, Malay), *buluh taris* (Malay).

Uses. In Sabah this species is not so common as in Brunei or Sarawak, where *balui* is used for making handicrafts and other items. In Lingkungan, however, culms of this species are used as fish stake, sailing masts and for framing.

Notes. The origin of this bamboo has not been identified as yet. It is possible that *balui* was introduced to the northern part of Borneo from southern Thailand, where a similar bamboo is found abundantly. Further study is needed to confirm the origin of *G. balui*.

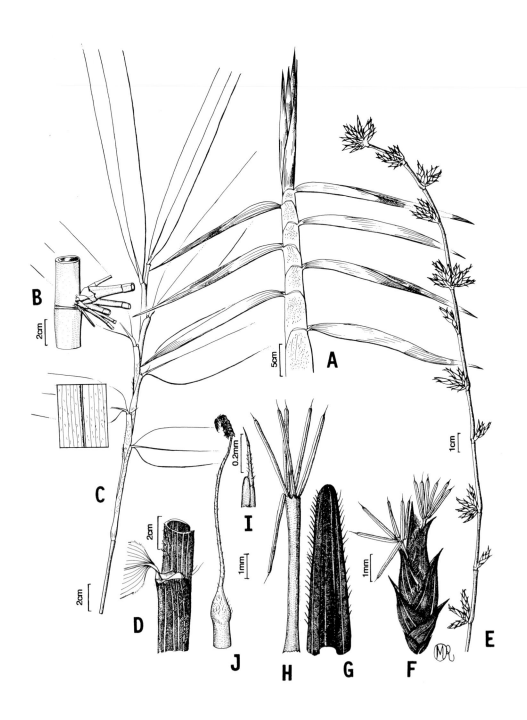

Fig. 13. *Gigantochloa balui*. A, Culm shoot. B, Mid-culm branch complement. C, Leafy branch. D, Leaf sheath, detail. E, Flowering branch. F, Pseudospikelet. G, Palea. H, Staminal tube with anthers. I, Anther apex, detail. J, Gynoecium. (A-D from fresh material; E-J from SAN 124361.)

2. **Gigantochloa levis** (Blanco) Merrill   (Fig. 14)

(Latin, smooth, probably referring to the lack of unevenness at the culm nodes)

Merrill in Amer. Journ. Bot. 3: 61 (1916); Holttum in Gard. Bull. Sing. 16: 119 (1958); Widjaja in Reinwardtia 10(3): 353 (1987); Wong in Sandakania 1: 15-12 (1992).

Synonyms:
Bambusa levis Blanco, Fl. Filipp. 1: 272 (1837).
Gigantochloa scribneriana Merrill in Phil. Journ. Sci. Bot. 1 (Suppl.): 390 (1906); Gamble in Phil. Journ. Sci. Bot. 5: 270 (1910).
Dendrocalamus curranii Gamble in Phil. Journ. Sci. Bot. 5: 271 (1910); Brown & Fisher in Bull. Phil. Bur. For. 22: 261(1920).

Culms 15-20(-30) m tall, diameter 5-14 cm, walls 1-1.2 cm thick, internodes 30-40 cm long, covered by brown hairs when young, becoming glabrous by age, light green to greyish green (because of lichen growth), nodes not swollen, lowermost nodes bearing short aerial roots; young shoots covered with dark brown hairs, often tinged with orange. Branch complement of few branches, short, hairy when young. Culm sheaths covered with dark brown hairs, 20-38 cm long, up to 48 cm wide at the base, narrowing to the apex, about 10 cm wide; blades 12-18 x 2-5 cm, deflexed, lanceolate to ovate lanceolate, tapering, densely hairy abaxially; ligule 15 mm tall, deeply laciniate, often with bristles along the edge; auricles large, erect or deflexed, up to 3 cm long, with long bristles along the edge (bristles about 15 mm long). Leaf blades glabrous or pubescent (especially when young), 18-32 x 2.5-5 cm, base attenuate or rounded, petiole up to 1 cm long; sheaths glabrous or with scattered appressed pale hairs, margins hairy (hairs easily detached); auricles small, often with bristles along the edge; ligule very short. Inflorescences borne on leafless branches, axes pubescent. Spikelets about 1 cm long, pubescent, consisting of 2-3 glumes, 3-4 florets, one empty lemma; glumes broadly ovate, with pointed tips, pubescent; lemmas pubescent, with pale hairs along the margins, with pointed tips; paleas hairy along the keels.

Distribution. So far this species has been recorded for Borneo and Luzon, planted, naturalized or growing spontaneously; it is possible that *G. levis* is also found in the eastern part of Indonesia.

Habitat. In Sabah this species is commonly found on slopes of hills or in gullies above 200 m altitude.

Vernacular name. *Poring* (Kadazan, Dusun, and generally).

Uses. General purposes, especially as building material; the *rebung* is very good to eat.

Notes. The original description of *G. levis* (first named *Bambusa levis*) was based on a plant found in Luzon (the Philippines). Holttum also identified one

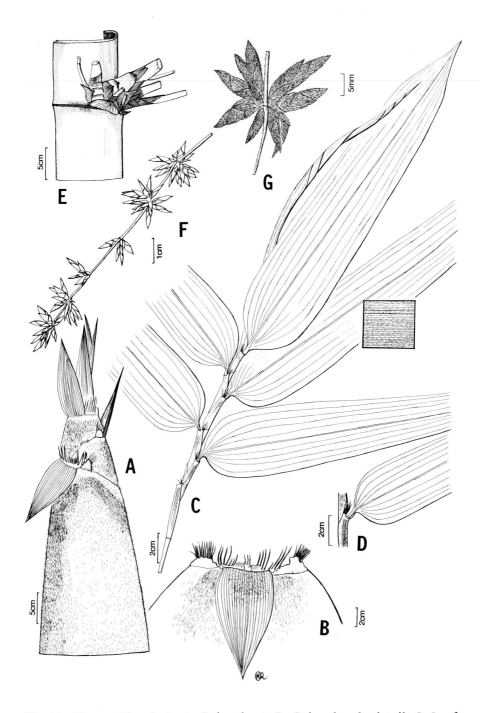

Fig. 14. *Gigantochloa levis*. A, Culm shoot. B, Culm sheath, detail. C, Leafy branch (with inset showing magnified view of lower leaf surface). D, Leaf sheath, detail. E, Mid-culm branch complement. F, Flowering branch. G, Cluster of pseudospikelets. (A, B from SAN 85963; C, D from SAN 42601; E from fresh material; F, G from SAN 85900.)

of the large bamboos in Peninsular Malaysia as *G. levis*. Wong (1992) has shown that the bamboo from Peninsular Malaysia is slightly different from *poring* in Sabah (or Sarawak and Brunei), and is a distinct species, *G. thoii*.

Because of its large-size culms, *poring* is often misidentified with *Dendrocalamus asper*, a large bamboo commonly known as *betung*. *D. asper* can be distinguished from *poring* by its young shoots covered with black hairs, the very short lowermost internodes covered by golden brown hairs and the swollen lowermost nodes bearing dense aerial roots.

RACEMOBAMBOS Holttum

(Latin: *racemus*, flowering branch resembling a raceme, referring to the arrangement of the spikelets)

Holttum in Gard. Bull. Sing. 15: 267-273 (1956), Kew Bull. 21: 281 (1967); Dransfield in Kew Bull. 37(4): 661-679 (1983).

Loosely tufted, scrambling bamboos; culms glabrous, hollow, usually thin walled, straight, internodes terete, nodes not swollen. Branch complement at mid-culm nodes of several to many branches, the primary branch dominant and elongate, secondary branches smaller bearing leaf blades. Culm sheaths thin, papery when dry, usually glabrous, narrowing upwards, eventually deciduous; auricles small, in some species with long bristles; blades erect or deflexed usually glabrous. Leaf blades variable in size, from 4-6 cm long (*R. gibbsiae*) to 15 cm long (*R. hirsuta*); ligule very short; auricles short with long bristles. Inflorescences raceme-like, semelauctant, 2-7 cm long, terminating leafy branches, axis glabrous or hairy. Spikelets 10-50 mm long, glabrous or hairy, comprising 2-3 glumes, 3-8 perfect florets, one rudimentary terminal floret; lowermost glume small, two upper glumes longer, with several nerves; rachilla internodes glabrous or hairy; lemma glabrous or hairy, 5-9-nerved; palea 2-keeled, glabrous or hairy in the upper part; lodicules 3; ovary usually hairy on the upper part, stigmas 3; stamens 6.

Distribution. Peninsular Malaysia, Borneo, Palawan, Sulawesi, Seram, New Guinea, New Ireland, Solomon Islands. There are 17 species in *Racemobambos*, and six are found in Sabah. *R. gibbsiae* is endemic to Mt. Kinabalu at the altitude 2000-3000 m, *R. rigidifolia* at Marai Parai (on Mt. Kinabalu). *R. hepburnii* is endemic to Sabah, known only from Mt. Kinabalu at 1100-2000m and also from the Crocker Range. *R. pairinii* is known from only a few localities on ultramafic substrate and is probably also endemic to Sabah.

Habitat. Montane forest, except *R. pairinii* which is also found in small-crowned forest at low altitude (*c*. 50 m) on ultramafic soil.

Notes. The genus *Racemobambos* is related to *Arundinaria* by having semelauctant inflorescences in the form of racemes bearing multiflorous spikelets. The genus *Arundinaria* and many of its relatives are native to temperate or cold countries.

The branch system is very characteristic in *Racemobambos*, except that of *R. rigidifolia* and *R. hirsuta* (also *R. setifera* from Peninsular Malaysia). It is almost always possible to recognise the genus even without flowering or fruiting material, as long as the developing vegetative branches are available. At each node of the culm there is a single branch bud enclosed by a prophyll. Inside the prophyll there is a row of small branch buds, arranged on each side of a dominant and larger middle branch, as if all the branches are borne on a common base. The middle branch bud will develop into a dominant branch which has the ability to produce branches at each node in the manner of the main culm bearing it. The smaller branch buds, 3-9 in each group, will develop into subequal short branches which are erect first, but later spreading geniculately. They usually do not branch again, but bear leaf blades and may also terminate·in inflorescences.

Key to the species

1.   Branches up to five at each (mid)culm node, middle branch usually not
       dominant; spikelets usually with two glumes 2
       Branches 7-19 at each (mid)culm node, middle branch dominant and
       elongating; spikelets usually with three glumes 4

2.   Spikelet glabrous .... *R. rigidifolia*
       Spikelet hairy 3

3.   Top of culm sheath horizontal, ligule very short, auricles small
       with long bristles; leaf ligule short, auricle with bristles .... *R. hirsuta*
       Top of culm sheath recessed towards the middle, ligule about 1 mm
       high, auricles not present and without bristles; leaf ligule elongate and
       obtuse, without bristles .... *R. pairinii*

4.   Palea hairy (found only on Mt. Kinabalu, above 2000m) .... *R. gibbsiae*
       Palea glabrous (on Mt. Kinabalu and other mountains, below 2000m) ...
       5

5.   Spikelet not more than 12 mm long, 4-flowered; leaf blades narrow,
       2.5 mm wide (Gunung Meligan; Sarawak & Brunei) .... *R. glabra*
       Spikelet over 15 mm long, 5-8-flowered; leaf blades 4-8 mm wide (Mt.
       Kinabalu and the Crocker Range) .... *R. hepburnii*

1. **Racemobambos gibbsiae** (Stapf) Holtt. (Fig. 15)

(named after Miss L. S. Gibbs, a plant collector credited with the first recorded ascent of Mt. Kinabalu by a woman)

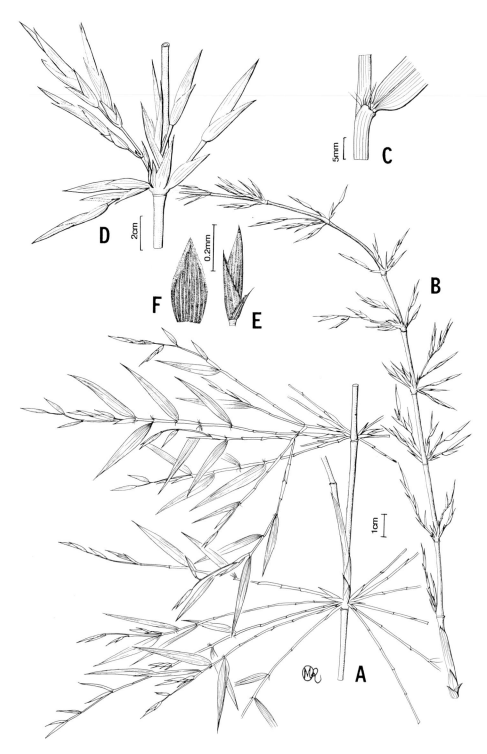

Fig. 15. *Racemobambos gibbsiae*. A, Mid-culm branch complement, flowering. B, Flowering branch. C, Leaf sheath, detail. D, Cluster of spikelet-bearing branches. E, Glumes. F, Lemma. (All from SD 757.)

Holttum in Gard. Bull. Sing. 15: 272 (1956).

Synonym:
Bambusa gibbsiae Stapf in Journ. Linn. Soc. Bot. 42: 189 (1914).

Culms slender, first erect then leaning or scrambling over nearby vegetation; internodes 25-40 cm long, diameter 1 cm, smooth. Branches many at each node, middle branch dominant and elongating, the secondary branches short, straight or geniculate, bearing leaf blades often with an inflorescence terminating it. Culm sheaths smooth, glabrous, purplish; auricles small, with bristles; ligule very short; blades narrow, erect. Leaf blades 4-6 x 0.4 cm, lanceolate, glabrous; sheaths minutely pubescent; auricles small, bearing long bristles. Inflorescence a short raceme, up to 5 cm long, bearing 5-7 spikelets, axes glabrous, slender; spikelet 20 mm long, 2(3)-flowered with a reduced terminal floret; rachilla internodes 7 mm long, slender; glumes 3; lemma 10 x 3 mm, hairy especially in the upper part; palea 8 mm long, hairy on the back.

Distribution. *R. gibbsiae* is only found on Mt. Kinabalu at 2000-3000m.

Habitat. Montane forest above 2000 m.

Notes. The type of *R. gibbsiae* was collected at Marai Parai by Miss Gibbs; this species is also common along the mountain tracks from Kamborangoh to Layang-Layang. The plants are either scrambling over nearby trees or forming a bushy growth of entangled culms which cannot support themselves. Young shoots are erect and often seen sticking up among the entangled mature culms. *R. gibbsiae* has a gregarious flowering habit. Flowering specimens were collected in 1910, 1933, 1957, and 1967. In 1979 the whole population along the mountain tracks produced flowers (*S. Dransfield* SD755 and SD759), and continued to flower in 1981 when it was observed (K.M. Wong). In 1986 the culms that had flowered had disappeared and were replaced by young and mature sterile culms (Wong *et al.* 1988). Since then there has been no record of its further flowering.

2. **Racemobambos glabra** Holtt. (Fig. 16)

(Latin: *glabra*, hairless; referring to the spikelets)

Holttum in Gard. Bull. Sing. 15: 270 (1956); Kulip in Sandakania 1: 7-9 (1992).

Culm 1.2 cm diameter, thin-walled, internodes 25-30 cm long, smooth, glabrous, white-waxy below nodes, nodes not swollen. Culm sheaths glabrous; blades narrowly lanceolate, erect or spreading, glabrous; ligule short, entire; auricles not conspicuous, with short white bristles. Leaf blades 7-8 cm long, up to 5 mm wide smooth, glabrous, sheath glabrous, ligule short, auricles not conspicuous but with long white bristles. Inflorescence 2-3 cm long, racemose, comprising 3-4 spikelets; spikelets 1-1.7 cm long, 4-5-flowered, 3 glumes, floret about 5 mm long, lemma glabrous, palea glabrous.

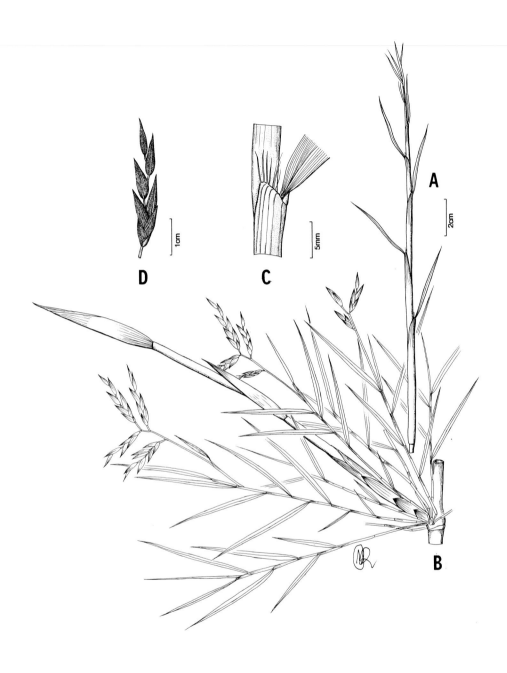

Fig. 16. *Racemobambos glabra*. A, Culm shoot. B, Mid-culm branch complement, with leafy and flowering branches. C, Leaf sheath, detail. D, Spikelet. (A, C from SAN 132701; B, D from WKM 773.)

Distribution. Sabah: Sipitang (Meligan Forest Reserve); Sarawak: Gunung Mulu. Brunei: Retak and Pagon ridges.

Habitat. Mountain forests.

Notes. This species was collected recently by J. Kulip in sterile state and is a new record for Sabah (Kulip 1992). It is related to *R. hepburnii*, but differs from it by its narrow leaf blades and smaller spikelets and florets.

3. **Racemobambos hepburnii** S. Dransf. (Fig. 17)

(named after A. John Hepburn, a forester long-time serving in Sabah)

Dransfield in Kew Bull. 37(4): 670 (1983).

Culms scrambling or leaning on nearby trees, internodes 25 cm long, diameter 8-9 mm. Branch complement 10-29 at each node, middle branch dominant and elongate. Culm sheath about 17.5 cm long, glabrous, smooth, narrower upwards; auricles not conspicuous, with bristles; blades narrow, usually deflexed, easily shed; ligule very short, ciliate. Leafy branches 10-15 cm long; leaf blades 6-12 x 0.5 cm, glabrous; sheath glabrous or pubescent, hairy along the margins; auricles not conspicuous, with bristles; ligules short. Inflorescences 13-25 cm long, bearing 3-7 spikelets, main axis and lateral branches usually glabrous; spikelet 3-5 cm long, with 5-8 florets; rachilla internodes 5 mm long, glabrous; glumes 3; lemma 9 x 5 mm, glabrous; palea 8-9 x 3 mm, glabrous.

Distribution. Mt. Kinabalu, at 1100 to 2000 (2200) m, and also in the Crocker Range.

Habitat. Lower montane forest below 2000 m.

Notes. *R. hepburnii* is closely related to *R. gibbsiae* and *R. glabra*. In *R. hepburnii* the spikelets are glabrous and are larger that those of *R. glabra*. This species is found abundantly in forest around the Park Headquaters, Kinabalu National Park. In 1979 all plants were sterile, and no flowers were found. In 1986 the population there produced flowers and probably continued flowering until 1988 when it was observed last (Wong *et al.* 1988). Like *R. gibbsiae*, *R. hepburnii* has also a gregarious flowering habit.

4. **Racemobambos hirsuta** Holtt. (Fig. 18)

(Latin, hirsute, referring to the hairy spikelets)

Holttum in Gard. Bull. Sing. 15: 272 (1956).

Open tufted bamboo 3-5 m tall, or long scrambling bamboo. Culms 5-10 mm diameter, smooth, internodes 40-60 cm long. Culm sheaths glabrous,

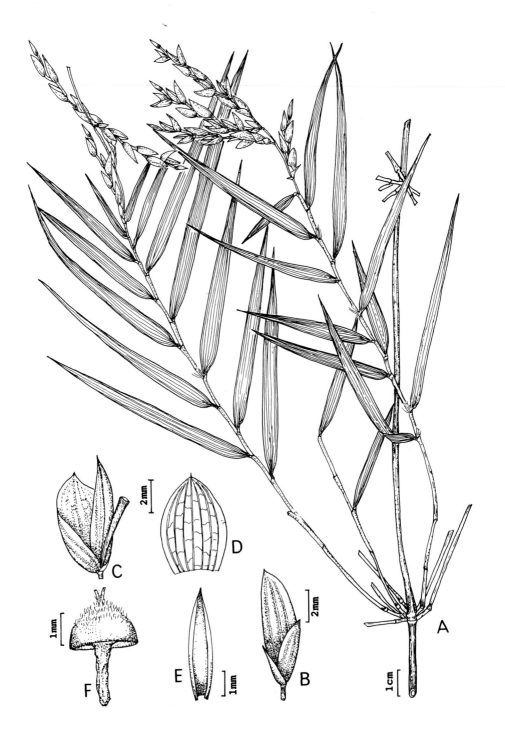

Fig. 17. *Racemobambos hepburnii*. A, Flowering branches. B, Glumes. C, Floret and rachilla internode. D, Lemma. E, Palea. F, Ovary (style removed). (All from RSNB 4111.)

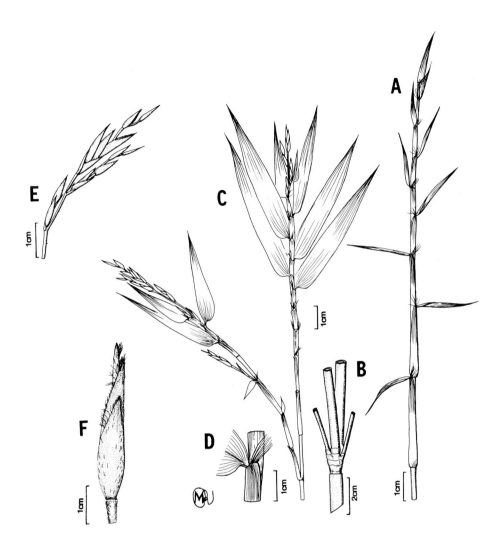

Fig. 18. *Racemobambos hirsuta*. A, Culm shoot. B, Mid-culm branch complement. C, Flowering leafy branches. D, Leaf sheath, detail. E, Spikelets. F, Detail of spikelet. (All from WKM 2107.)

51

margins hairy, top part horizontal, light green; blade narrow, 7-8 x 0.5 cm, deflexed, glabrous; ligule very short; auricles not seen. Leafy branches 30-40 cm long, with 6-8 leaf blades; leaf blades 15-25 x 3-4 cm, glabrous; ligule entire, 3 mm long; auricles small with long bristles. Inflorescence about 15 cm long, bearing 4-5 spikelets; spikelet 5-6 cm long, 5-6-flowered; glumes 2; rachilla internodes about 1 cm long, pubescent; lemma hairy, 1.2 cm long; palea shorter than lemma, hairy especially along the keels.

Distribution. Sabah: Penibukan on Mt. Kinabalu, Bukit Silam, Mt. Nicola.

Habitat. Forest on ultramafic soil, from 50 m up to 1500 m.

Notes. *R. hirsuta* is very common at Penibukan and Bukit Silam.

5. **Racemobambos pairinii** K.M. Wong (Fig. 19)

(named after Datuk Sri Joseph Pairin Kitingan, Sabah's Chief Minister)

Wong in Sandakania 1: 1-6 (1992).

Open tufted bamboo. Culms 3-5 m long, erect with pendulous tips, or scrambling and flopping over surrounding vegetation, 5-9 mm in diameter, internodes 30-40 cm long, glabrous. Branch complement with a dominant middle branch and 2-5 secondary branches. Culm sheaths glabrous, middle of top recessed at attachment of blade; blades linear, spreading or reflexed; ligule a subentire rim; auricles not developed. Leaf blades 7-24 X 1.5-3 cm, glabrous; sheaths glabrous; ligule elongate, obtuse, up to 4 mm long, entire; auricles not developed. Inflorescence 7-11 cm long, bearing 4-5 spikelets, main axis pubescent; spikelet 4.5-6 cm long, with 2 glumes and 4-6 perfect florets; rachilla internodes about 7 mm long, minutely hairy; lower glume 4-8 mm long, upper glume up to 14 mm long, pubescent; lemma 14-16 mm long, mucronate, coriaceous, pubescent; palea shorter than the lemma, minutely hairy on the back and along the keels.

Distribution. Sabah: Karamuak Valley, Bukit Silam, Mt. Nicola in Danum Valley.

Habitat. Small-crowned forest on ultramafic soil.

6. **Racemobambos rigidifolia** Holtt. (Fig. 20)

(Latin, rigid or stiff leaf blades)

Holttum in Gard. Bull. Sing. 15: 273 (1956).

Culms 5-7 mm diameter, green to yellowish green, smooth, internodes 40-50 cm long. Branch complement 3-4 at each node, more or less of the same

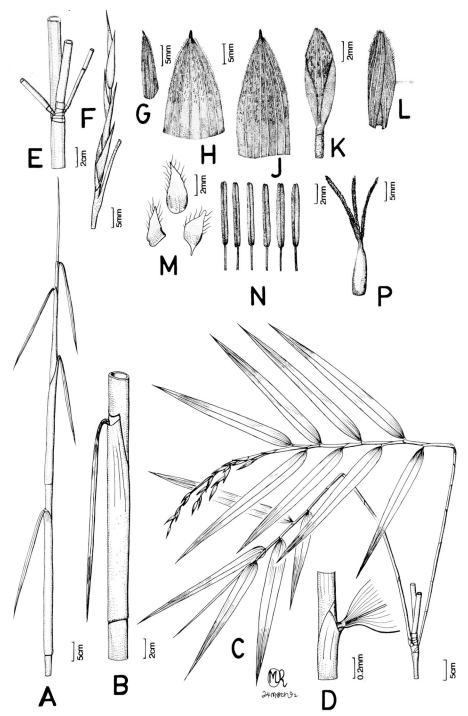

Fig. 19. *Racemobambos pairinii*. A, Culm shoot. B, Culm sheath. C, Flowering and leafy branches. D, Leaf sheath, detail. E, Mid-culm branch complement. F, Spikelet. G, Lower glume. H, Upper glume. J, Lemma. K, Palea, back view. L, Palea, front view. M, Lodicules. N, Stamens. P, Gynoecium. (All from WKM 2106.)

length, erect. Culm sheath light green, glabrous; auricles not conspicuous, with short bristles; blades narrow, about 15 cm long, deflexed; ligule short with bristles; auricles not present. Leafy branches 30-60 cm long, bearing 8-18 leaf blades; leaf blades 20-25 x 3-4 cm, usually glabrous, rigid; ligule short; auricles not conspicuous, but with long bristles. Inflorescence about 20 cm long, bearing 2-3 spikelets; spikelet 30 mm long, 4-flowered, glabrous; glumes 2; rachilla internodes 8 mm long, glabrous; lemma about 15 mm long, glabrous; palea 10 mm long, glabrous.

Distribution. So far found only at Marai Parai, Mt. Kinabalu.

Habitat. Montane forest above 1600 m, on ultramafic soil.

Notes. *R. rigidifolia* and *R. hirsuta* are closely related species. The former differs from the latter in having glabrous spikelets and rigid smaller leaf blades. *R. rigidifolia* is found above 1600 m at Marai Parai, whereas *R. hirsuta* is found below 1600 m at Penibukan on Mt. Kinabalu, and also in forests on Bukit Silam, at 50-800 m. It seems that there is a clear-cut boundary in altitude between these two species.

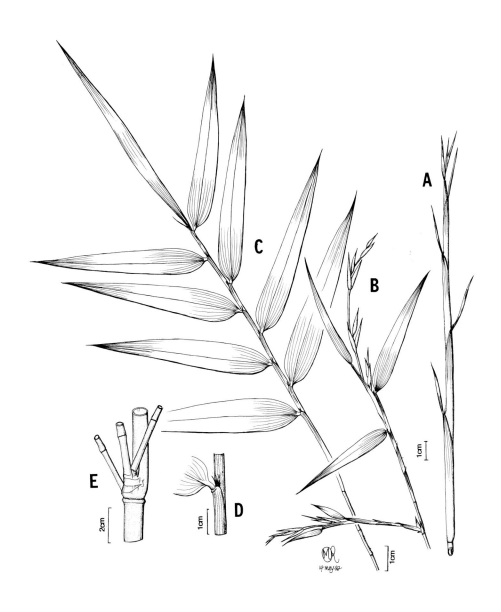

Fig. 20. *Racemobambos rigidifolia.* A, Culm shoot. B, Flowering branch. C, Leafy branch. D, Leaf sheath, detail. E, Mid-culm branch complement. (All from SD 715.)

SCHIZOSTACHYUM Nees

(Greek: *schistos*, cleft, divided; *stachys*, spike; alluding to the spaced grouping of the spikelets)

Nees, Agrost. Bras. 534 (1829); McClure in Blumea 2: 86-97 (1936); Holttum in Gard. Bull. Sing. 16: 31 (1958); Monod de Froideville, in Backer & Bakhuizen, Flora of Java 3: 640 (1968); Dransfield in Kew Bull. 38(2): 321-331 (1983).

Closely tufted bamboo. Culms erect with drooping tips, or somewhat scrambling, thin-walled, green to yellowish green (or yellow), often hairy when young (hairs usually pale brown or white), becoming glabrous; internodes 40-100 cm long, nodes usually not swollen. Young shoots covered with golden brown, white or pale brown hairs. Branch complement of many branches at each node from midculm upwards, short, more or less of the same length, young branches developing almost simultaneously in the culm bud (except in *S. terminale*, where the middle branch is dominant and often elongate). Culm sheaths usually hairy, hairs white to red brown (rarely dark brown); blades erect or deflexed, linear or broadly triangular, glabrous or hairy; auricles present or absent; ligule short and entire, or long and toothed. Leaf blades usually with conspicuous leaf stalks, glabrous or hairy (especially when young); auricles present or absent; ligule short or long. Inflorescences usually terminating leafy branches, about 40 cm long. Spikelets slender and cylindrical, or long and flattened, glumes usually absent, containing one to several florets, and a rachilla extension bearing a rudimentary floret; lemmas acuminate or with long-pointed tip, glabrous or hairy, many-nerved; paleas of multiflowered spikelets 2-keeled, that of one-flowered spikelets not keeled but with a long groove on the back, and without or with 2-pointed tips, hairy or glabrous; lodicules present or absent; stamens six; ovary long and slender; fruit with moderately thick and hard pericarp.

Distribution. From Thailand, throughout Malesia, to the Pacific Islands. There are about 45 species in the genus. In Sabah there are seven species, and one of them is endemic.

Habitat. Bamboos of the lowlands, wild in forest, or spontaneous in wastelands and roadsides; sometimes planted.

Uses. Because *Schizostachyum* species are readily available (found growing wild or spontaneously) and have culms with thin walls, many of the species are widely used by local people. *S. brachycladum* and *S. zollingeri* have large-diameter culms (up to 10 cm) which are relatively light. The culms are used for making rafts, flooring or roofing, water containers, and many other uses. Other species with small-diameter culms are used for making musical instruments and other items (such as blow-pipes, tobacco containers etc.).

Notes. The genus in general is easily recognised in the field by the thin-walled culms with many short branches at each node, the conspicuously stalked leaf blades, and the young shoots which are usually covered by pale hairs. Most of the species produce flowers continuously.

Key to the species

1. Middle branch of branch complement dominant in size, either dormant or elongate .... *S. terminale*
    Middle branch of branch complement not dominant, developing into the same size as secondary branches 2

2. Culms 5-10 cm diameter; blades of culm sheaths erect 3
    Culms up to 2 cm diameter; blades of culm sheaths deflexed 4

3. Culms erect with drooping tips, green to bluish green, or somewhat yellow; culm sheaths with red-brown hairs on the back; blades of culm sheaths broadly triangular, 7 cm long; auricles of culm sheath prominent with long bristles .... *S. brachycladum*
    Culms erect when young, then leaning onto trees or drooping to the ground, light green; culm sheath with light brown hairs on the back; blades of culm sheaths narrowly triangular, tapering to the tip, 20-25 cm long; auricles of culm sheath small without bristles .... *S. pilosum*

4. Blades of culm sheaths lanceolate, long, tapering; auricles of culm sheath not prominent or absent 5
    Blades of culm sheaths narrowly triangular or ovate lanceolate; auricles of culm sheath prominent with long bristles 6

5. Ligule of culm sheath not conspicuous, but with long bristles; top of culm sheath recessed in the middle .... *S. lima*
    Ligule of culm sheath conspicuous, up to 2 mm long, with short bristles ; top of culm sheath more or less horizontal ....
    .... *Schizostachyum* sp.

6. Culms generally drooping; culm sheath with white hairs; blades of culm sheaths 7 mm wide; leaf blades dark green .... *S. blumei*
    Culms erect with drooping tips; culm sheath with light brown hairs; blades of culm sheath 15 mm wide; leaf blades light green ....
    .... *S. latifolium*

1. **Schizostachyum blumei** Nees   (Fig. 21)

(named after C.L. Blume, a Dutch botanist)

Nees, Agrost. Bras. 534 (1829); McClure in Blumea 2: 86 (1936); Dransfield in Kew Bull. 32(2): 330 (1983).

Synonyms:
Melocanna longispiculata Kurz, in Teijsmann & Binnendijk, Cat. Hort. Bot. Buitz. 20 (1866) *(nom. nud.)*.
M. zollingeri Steud. var. *longispiculata* Kurz ex Munro in Trans. Linn. Soc. 26: 143 (1866).
Schizostachyum longispiculatum (Kurz ex Munro) Kurz in Journ. As. Soc. Beng. 39(2): 89 (1870).

Culms erect when young, later drooping or arching, about 7 m long, diameter 1-2 cm, dark green, rough when young, becoming smooth, internodes 40 cm long, with white wax below nodes; young shoots light green. Culm sheaths light green with appressed white hairs, 15-15.5 cm long; blades erect first then deflexed, 13 x 0.7 cm, hairy adaxially especially near the base; auricles 8 mm long, 1 mm tall with long bristles along the edge (bristles up to 11 mm long); ligule very short with long bristles along the edge. Leaf blades 20-35 x 5-6.5 cm, base more or less rounded, glabrous, dark or dull green; sheath glabrous; auricles prominent, 1 mm long with long bristles along the edge (bristles up to 13 mm long); ligule very short. Inflorescence about 20 cm long. Spikelets up to 25 mm long, one-flowered with a rachilla extension bearing a rudimentary floret; lemma with a long-pointed tip, hairy near the apex; palea with 2 long-pointed tips, hairy near the apex; lodicules not present.

Distribution. Borneo (Sabah, Sarawak; Brunei; Kalimantan), introduced in Java.

Habitat. Lowland and hill dipterocarp forests, forest margins, or on wastelands by roadsides.

Local name. *Tombotuon* (Kadazan).

Notes. *S. blumei* was described based on a specimen (the type specimen) collected from a plant growing in Bogor, Indonesia, which was probably introduced from Borneo. The type specimen was in the Berlin herbarium, but was destroyed during the war. Fortunately, McClure (1936) saw the type and represented it with a complete drawing. The drawing matches the plant cultivated in the Bogor Botanic Garden, which is also the living type specimen of *S. longispiculatum*. *S. blumei* is related to *S. latifolium*, but in the field they can be distinguished from one another by their habits. *S. latifolium* is usually a slender erect bamboo; the clump is usually open, the culms are light green with drooping tips; the leaf blades are also light green, and the culm sheaths have prominent auricles that do not fall off in the mature culm. *S. blumei* is also a slender bamboo, but the culms are arching or leaning; the clump is closed or compact and in the form of a bush, with several young erect culms sticking out;

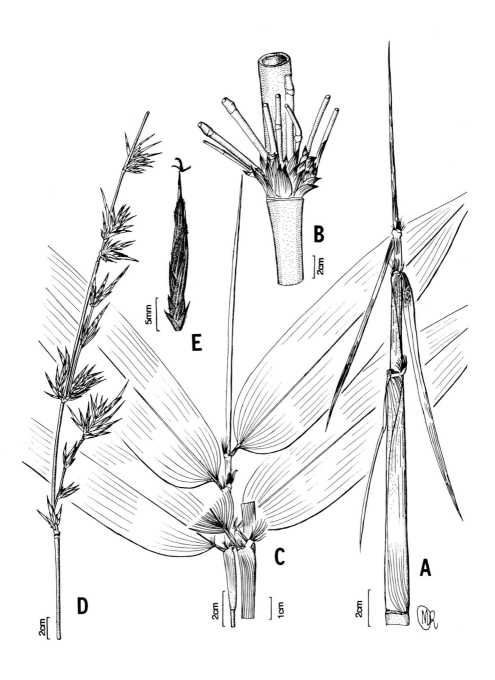

Fig. 21. *Schizostachyum blumei*. A, Culm shoot. B, Mid-culm branch complement. C, Leafy branch and detail of leaf sheath. D, Flowering branch. E, Pseudospikelet. (All from SD 740 except C, from A 3278.)

the culms are usually dark green, and the leaf blades are dark or dull green.
Morever, *S. blumei* differs from *S. latifolium* by the absence of lodicules in the
floret.

2. Schizostachyum brachycladum Kurz (Fig. 22)

(Greek, referring to the short branches)

Kurz in Journ. As. Soc. Bengal 39(2): 89 (1870); Holttum in Gard. Bull. Sing. 16:
45 (1958).

Culms 10-15 m tall, erect with drooping tips, green, bluish green or
yellow, internodes up to 20 cm long, diameter 5-10 cm; young shoots creamy
brown. Culm sheaths about 20 cm long, thick, rigid, creamy to yellowish brown,
covered with red-brown hairs especially when young; auricles 5-6 mm long, with
long bristles (bristles 8 mm long, easily shed); blades broadly triangular, erect,
rigid, 7-18 cm long, 7-9 cm wide at the base, glabrous; ligule short. Leaf blades
20-32 x 4-5.5 cm, stalks 9 mm long, glabrous or hairy abaxially, especially when
young, base rounded; sheaths hairy when young, glabrous with age; auricles
small with long bristles (bristles up to 7 mm long); ligule very short, entire.
Inflorescence 15 cm long. Spikelets about 12 mm long, usually glabrous,
containing 1-2 florets; lodicules 3.

Distribution. Native in Malesia, it is found all over the region.

Habitat. *S. brachycladum* is found growing spontaneously in wastelands, or
planted in villages. It is also found growing wild in disturbed or secondary
forests, on sandy clay or sandy loam soil, at 20-600 m.

Local names. This species is generally simply called *buloh* or *wuluh* in Sabah,
especially by the Kadazan people. However, the species is also referred to as
buluh lemang or *buluh nipis* in Malay.

Uses. This species is widely used by local people, especially for cooking glutinous
rice, called *lemang*, and also for water containers. It is also used for crafting
baskets or *wakid* (J. Kulip, pers. comm.).

Notes. The yellow variety of *S. brachycladum* is found commonly growing along
the road between Tambunan, Keningau and Nabawan, and is called *ru gading* by
the Dusun people. It seems that this variety is not planted in this area, but it is
not known with certainty whether it is native either. This variety has become a
popular ornamental plant in South East Asia.

In the field *buluh lemang* is easily recognised by its straight, erect,
smooth culms with thin walls, and creamy to yellowish brown culm sheaths with
erect, rigid, triangular blades. *S. brachycladum* is related to *S. zollingeri*, a species
common in southern Thailand, Peninsular Malaysia and Sumatra.

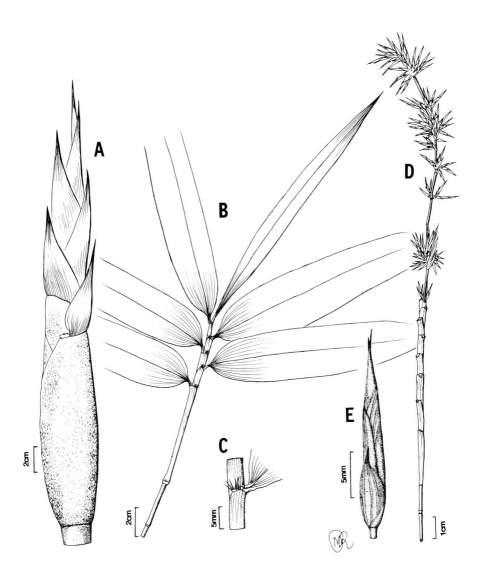

Fig. 22. *Schizostachyum brachycladum*. A, Culm shoot. B, Leafy branch. C, Leaf sheath, detail. D, Flowering branch. E, Pseudospikelet. (All from SD 721 except A, from SAN 86074.)

3. Schizostachyum latifolium Gamble (Fig. 23)

(Latin, large or broad leaf blade)

Gamble in Ann. Roy. Bot. Gard. Calcutta 7: 117 (1896); Holttum in Gard. Bull. Sing. 16: 48 (1958) (as *S. longispiculatum*); Dransfield in Kew Bull. 38(2): 331 (1983).

      Culms about 7 m tall, diameter 1.5-2 cm gradually reduced to 5 mm at the tip, erect with drooping tips, usually light green, hairy when young, becoming glabrous and smooth, internodes 60-80 cm long, whitish below nodes; young shoots light green. Culm sheaths green becoming yellowish brown when dry, long persistent on the culm, covered with light brown hairs, 10-13 cm long; blades erect first then deflexed, 8- 13 cm long, 15 mm wide, edges rolled towards the apex, hairy adaxially near the base; auricles prominent, 11 mm long, with long bristles (bristles 11 mm long); ligule short, entire. Leaf blades 12-30 x 3-5.5 cm, with stalks 6 mm long, base usually rounded, glabrous; auricles short, but with long bristles (bristles up to 12 mm long); ligule very short, entire. Inflorescence up to 27 cm long. Spikelets 2-3 cm long, one-flowered with a rachilla extension bearing a rudimentary floret: lemma hairy especially near the apex, with long tip; palea with 2 long tips, hairy in the upper part; lodicules 3.

Distribution. Peninsular Malaysia, Sabah; Sarawak. Indonesia: Kalimatan, Sumatra.

Habitat. Lowland dipterocarp forests, and wastelands by roadsides.

Local names. *Pelupu* (Dusun), *buluh lachau* (Iban).

Uses. In Sarawak the culms are used in basketry.

Notes. In Sabah this species is commonly found on wastelands around Papar, and also here and there around Poring. There are some clumps planted in the Agriculture Station, Ulu Dusun.

4. Schizostachyum lima (Blanco) Merrill (Fig. 24)

(Spanish: *lima*, file; culm surface rough as a file)

Merrill in Amer. Journ. Bot. 3: 62 (1919); Holttum in Kew Bull. 21: 278 (1967).

Synonyms:
Bambusa lima Blanco, Fl. Filip. ed.1, 271 (1837).
Schizostachyum hallieri Gamble in Philipp. Journ. Sci. Bot. 5: 274 (1910).

      Culms erect, straight, with drooping tips, 7-10 m tall, diameter 3 cm, internodes 30-100 cm long (or more), dark green and hairy when young, becoming glabrous; young shoots green. Culm sheaths to 30 cm long, covered

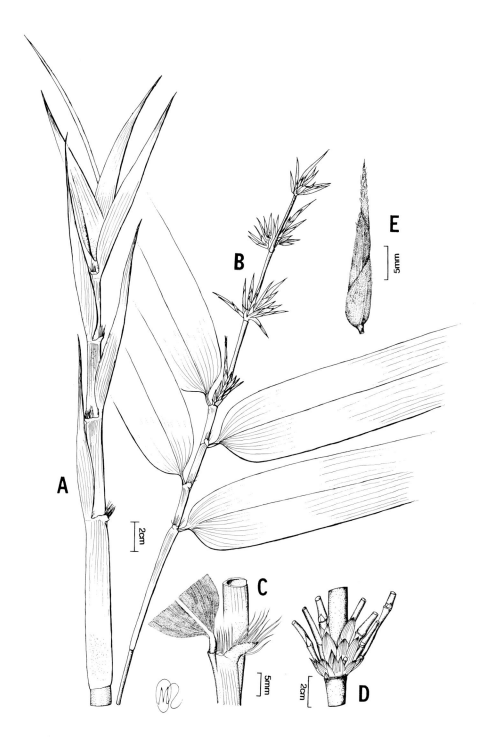

Fig. 23. *Schizostachyum latifolium*. A, Culm shoot. B, Flowering leafy branch. C, Leaf sheath, detail. D, Mid-culm branch complement. E, Pseudospikelet. (All from SD 722 except D, from SAN 124358.)

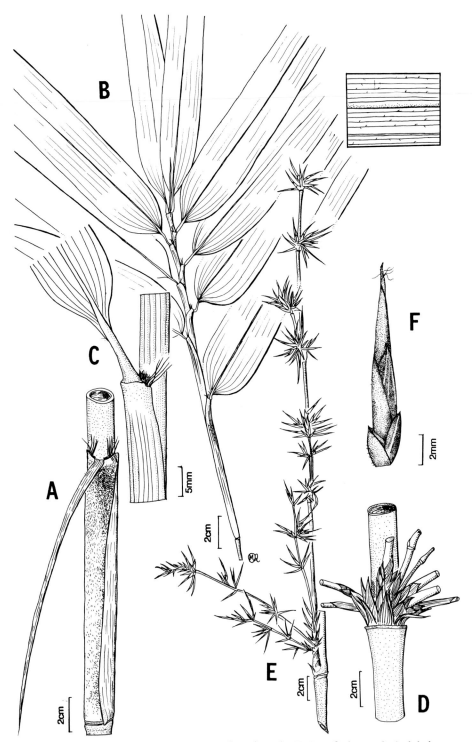

Fig. 24. *Schizostachyum lima*. A, Culm sheath. B, Leafy branch (with inset showing magnified view of lower leaf surface). C, Leaf sheath, detail. D, Mid-culm branch complement. E, Flowering branches. F, Pseudospikelet. (A-D from SD 753; E, F from A 5080.)

with brown to dark brown hairs, middle of top recessed at attachment of blade; blade erect first then deflexed, about 25 cm long, 5-6 mm wide, hairy adaxially near the base; ligule short with long bristles; auricles not prominent, but with long bristles. Leaf blades 19-22 x 3.5 cm, glabrous above, hairy below becoming glabrous; ligule short, irregularly toothed; auricles short with long bristles; sheaths glabrous. Inflorescence up to 20 cm long. Spikelet 12 mm long, glabrous, one-flowered with a rachilla extension bearing a rudimentary floret (sometimes absent); lemma acuminate; palea acuminate, not clearly 2-keeled.

Distribution. Sabah; Sarawak. Brunei. Philippines: Luzon. Indonesia: Kalimantan, Sulawesi, the Moluccas, Irian Jaya. Papua New Guinea.

Habitat. Lowland forests, forest margins, along rivers or river banks, and wastelands.

Local name. *Sumbiling* (Murut, Dusun).

Uses. Culms of this species are used as drinking tubes for consuming *tapai*, a local rice wine (J. Kulip, pers. comm.). Sometimes they are also used for the framework in roof-thatching.

Notes. *S. lima* is closely related to *S. iraten* from Java and to *S. jaculans* from Peninsular Malaysia. It differs from *S. iraten* in having a one-flowered spikelet, a culm sheath with a recessed middle portion at the top, and a culm sheath ligule with long bristles. *S. lima* differs from *S. jaculans* in having an acuminate palea, and a culm sheath with a recessed middle portion at the top. *S. jaculans* has a 2-pointed palea and a culm sheath with a truncate top. At present I consider this bamboo as *S. lima*, until a more critical study of the three species is carried out to determine whether they are conspecific or distinct.

5. **Schizostachyum pilosum** S. Dransf.   (Fig. 25)

(Latin, referring to hairs on the adaxial surface of the culm-sheath blade and the lower surface of the leaf blade)

Dransfield in Kew Bull. 38(2): 325 (1983).

Culms about 15 m long, diameter to about 5 cm at the base, erect when young, when mature leaning onto trees or drooping to the ground, internodes 30 cm long, covered with white wax and white or light brown hairs when young, becoming glabrous later, thin-walled; young shoots whitish (because of the presence of white wax). Culm sheaths about 30 cm long, covered with white wax and appressed white or light brown hairs, margins hairy; blades 30 x 4-5 cm, erect, stiff, tapering to the tip, edges rolled towards the apex, glabrous abaxially, adaxially densely covered with appressed white hairs; auricles about 1 mm long, with long bristles; ligule up to 8 mm long, irregularly toothed. Leaf blades 30-40 x 3.7 cm, base usually attenuate, densely hairy at the base on the lower surface, otherwise glabrous or with scattered appressed white hairs; stalk or petiole up to

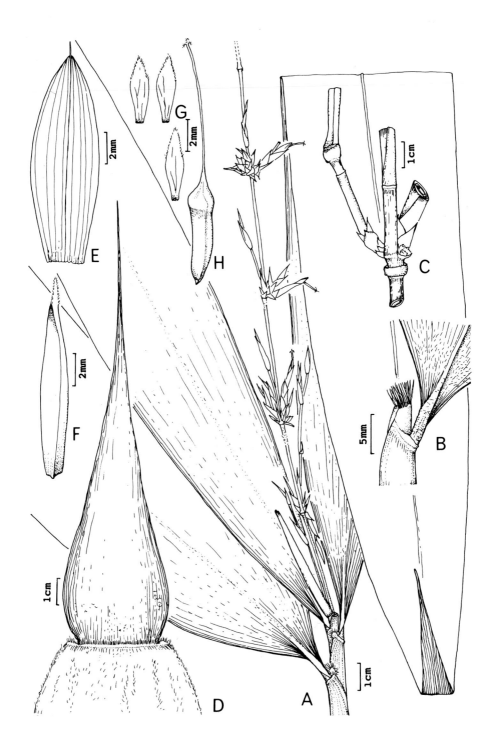

Fig. 25. *Schizostachyum pilosum*. A, Flowering branch. B, Lower part of leaf blade. C, Branches with swollen nodes. D, Culm sheath and blade. E, Lemma. F, Palea. G, Lodicules. H, Ovary. (A, E-H from Keith 1609; B-D from SD 745.)

1 cm long; auricles short with long bristles, easily detached; ligule about 5 mm long, fringed. Inflorescence usually with hairy axes, about 10 cm long. Spikelets to 2.4 cm long, consisting of 2 florets and a rachilla extension bearing a rudimentary floret; rachilla internodes 8 mm long; lodicules 3.

Distribution. So far found only in the area around Tambunan-Keningau-Nabawan, Sabah.

Habitat. Forest edges or margins, in damp places.

Local name. *Pus* (Murut, Dusun).

Uses. The culms are used in flooring and in basketry.

Notes. In the field this species can be recognised by its leaning habit with erect young culms. *S. pilosum* resembles *S. grande*, a species occurring in Peninsular Malaysia and southern Thailand, in its habit. A bamboo resembling *S. pilosum*, called *lempaki* by the Dusun people and used for making a wind instrument called *sompoton*, is found in the Interior District; the identification of this requires further investigation.

6. Schizostachyum terminale Holtt. (Fig. 26)

(Latin, alluding to the terminal inflorescence)

Holttum in Gard. Bull. Sing. 15: 274 (1956), 16: 51 (1958); Wong in Gard. Bull. Sing. 43: 39-42 (1991).

Open-tufted scrambling bamboo; culms 5-7 m long, diameter 8-15 mm, erect first then leaning on nearby trees, with drooping tips, not straight but slightly geniculate at the nodes, internodes 40-45 cm long, rough, hairy when young; young shoots slender, green. Branch complement of many branches with the middle branch dominant. Culm sheaths 9-21 cm long, green, with appressed white hairs when young, becoming glabrous with age; blades ovate-lanceolate to narrowly triangular, tapering to long tips, 6-11 cm long, up to 12 mm wide at the base, erect or slightly reflexed, densely hairy adaxially, remaining persistent on sheath; auricle small, developed only on one side. Leaf blades 13-21 x 2-5 cm, petiole 6 mm long, base slightly rounded, glabrous; auricles small with long bristles; ligule short, with relatively long bristles; sheath glabrous. Inflorescence 5 cm long, bearing few spikelets. Spikelet 25 mm long, glabrous, slender, terete; lemma and palea smooth, glabrous.

Distribution. Peninsular Malaysia; Sabah (found so far along the lower Kinabatangan River only). Brunei.

Habitat. Forest margins on flat, regularly flooded river banks.

Notes. This species is unusual in possessing a dominant middle branch, a feature not typical in *Schizostachyum* (Wong 1991). It has been recorded that the middle branch frequently remains dormant but later develops to reiterate the culm in size and length.

7. **Schizostachyum** sp., related to *S. iraten* Steud. from Java.

Culms erect with drooping tips, about 4(-10) m tall, diameter 1-1.3 cm, internodes covered with white hairs; a band of white waxy powder below the nodes, often with light brown hairs; young shoots light green with erect blades. Culm sheaths light green, covered with light brown hairs, 20-25 cm long, the top horizontal or slightly curved; auricles not prominent, or completely absent but replaced by long bristles; ligules 1-2 mm high, irregularly serrate, with short bristles; blades up to 32 cm long, up to 13 mm wide, narrowly lanceolate, tapering to fine tips, usually erect, often deflexed, white hairy adaxially. Leaf blades 19-28 x 2.5-4 cm, stalk (petiole) 1 cm long, glabrous, base attenuate; sheaths glabrous; auricles not present; ligule very short entire. Inflorescence 10-12 cm long, axes glabrous. Spikelets usually consisting of 2 florets, 15-25 mm long, glabrous; lemma glabrous, acuminate or with a short-pointed tip; palea glabrous, acuminate, 2- keeled; lodicules not present.

Distribution. So far this species is known only in Sabah and Sarawak.

Habitat. Hill forest, 100-300 m.

Local name. Also *sumbiling* (Murut), as for *S. lima*.

Notes. There are only a few collections of this species from Sabah and Sarawak. At present it is left unnamed until more collections become available and a more thorough study is carried out. It is closely related to *S. iraten* from Java, by having 2-flowered spikelets with acuminate paleas.

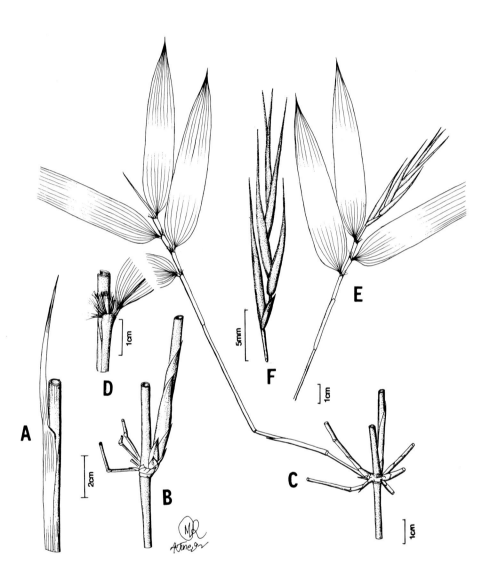

Fig. 26. *Schizostachyum terminale.* A, Culm sheath. B, Mid-culm branch complement. C, Leafy branch. D, Leaf sheath, detail. E, Flowering leafy branch. F, Inflorescence. (All from FRI 35151.)

SPHAEROBAMBOS S. Dransfield

(Greek: *sphaero*, globose, round; bamboo with globose fruit)

Dransfield in Kew Bull. 44(3) : 428 (1989).

Sympodial, erect or scandent bamboos. Culms straight or slightly zig-zag, hollow, usually with relatively thin walls. Branches few to many at each node, with primary branch dominant especially at the lower nodes. Inflorescence iterauctant, borne on leafless or leafy culms. Pseudospikelets few in a group at each node of the flowering branches; spikelets comprising of 3-5 florets, more or less laterally compressed; rachilla internodes slender; palea in mature spikelet longer than the lemma, 2-keeled, keels with wings. Mature fruits globose, smooth, pericarp relatively thick.

Distribution. So far the genus consists of three species, one species in Sabah, one in Mindanao (the Philippines) and one in North Sulawesi (Indonesia).

Sphaerobambos hirsuta S. Dransf.   (Figs. 27, 28, 29)

(Latin: *hirsutus*, hairy; young shoot and culm sheaths hairy)

Dransfield in Kew Bull. 44(3): 428 (1989).

Open-tufted bamboo; culms about 10 m long, diameter 3-4 cm, erect, then leaning (scrambling) or drooping to the ground, slightly zig-zag, internodes up to 40 cm long, with relatively thin walls, rough and hairy when young, smooth and glabrous later, nodes not swollen; young shoot green with white wax and stiff pale brown hairs. Branch complement of few to many branches at each node, with the middle branch dominant especially at the lower nodes, those at the upper nodes not so. Culm sheaths up to 20 cm long, rough and densely hairy when young, becoming glabrous, hairs stiff and irritant, easily shed; blades erect at first then deflexed or spreading, 10-25 cm long, about 6 cm wide at the base, tapering, hairy adaxially especially near the base; ligule very short, entire; auricles large, spreading, 7 mm tall, 2-2.5 cm long, with long bristles along the edge (bristles 6-7 mm long). Leaf blades 11-23 x 3-4 cm, base rounded, pubescent on both surfaces, densely hairy near the base and on the petiole; sheaths densely hairy; auricles 5 mm long, with bristles to 6 mm long; ligule short, laciniate. Inflorescences borne on leafless branches of leafy or leafless culms, or terminating leafy branches, axes hairy. Spikelets about 15 mm long, laterally flattened, consisting of 3 glumes and 5 florets (the uppermost much reduced); glumes pubescent; lemmas glabrescent; paleas longer than the lemma in mature spikelet, winged along the keels, hairy on the back especially between the keels; lodicules not present. Mature fruits globose, about 6 mm in diameter.

Distribution. So far this species has been found only around the Lohan village and river, between Ranau and Poring.

Habitat. Forest margin on ultramafic soil.

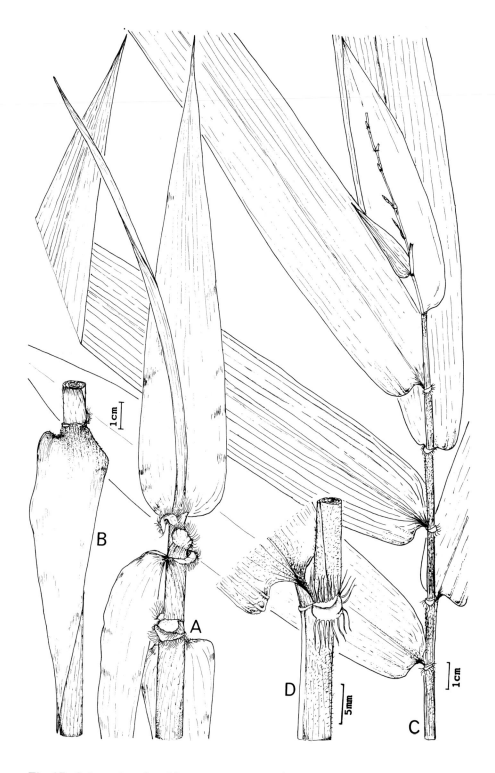

Fig. 27. *Sphaerobambos hirsuta.* A, Young shoot. B, Culm sheath. C, Leafy branch. D, Base of leaf blade. (All from SD 843.)

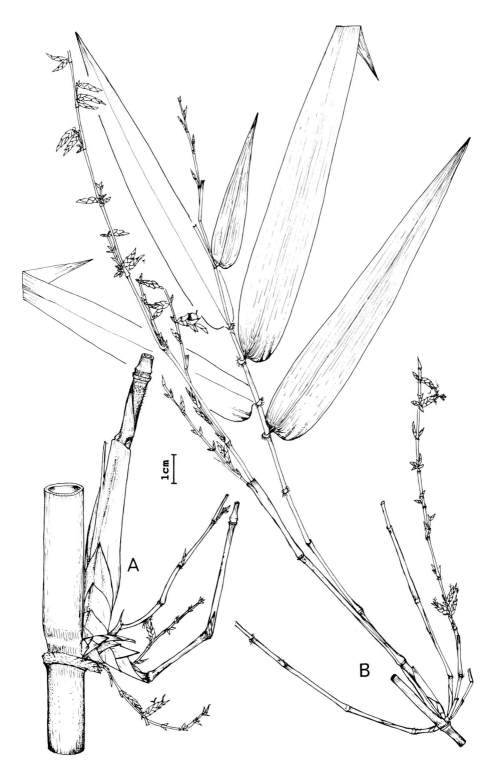

Fig. 28. *Sphaerobambos hirsuta*. A, Branch complement. B, Leafy and flowering branches. (All from SD 844.)

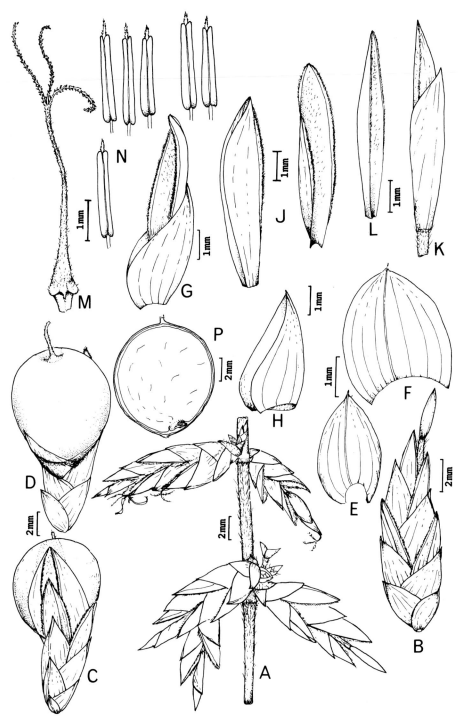

Fig. 29. *Sphaerobambos hirsuta*. A, Part of flowering branch. B, Spikelet. C-D, Spikelet with a fruit. E, Glume II. F, Glume III. G, Floret. H, Lemma. J, Palea (two views). K, Uppermost floret. L, Palea of uppermost floret. M, Ovary. N, Stamens. P, Section of fruit. (All from SD 844.)

THYRSOSTACHYS Gamble

(Greek: *thyrsos*, a thyrse, a particular type of inflorescence; *stachys*, the inflorescence; alluding to the disposition of the inflorescence)

Gamble in Ind. For. 20 : 1 (1894); Holttum in Gard. Bull. Sing. 16 : 80 (1958).

Tufted bamboo of a medium size. Culms erect, straight, solid or with a small lumen. Branches many at each node, borne from the midculm upwards. Culm sheath green, becoming stramineous with age, often persistent. Leaf blades small and narrow. Inflorescence iterauctant, borne on short leafless branchlets. Pseudospikelets in small groups on nodes subtended by a persistent short sheath.

Distribution. There are two closely related species in *Thyrsostachys*, both are native to Thailand and its neighbouring Indochinese countries.

Thyrsostachys siamensis Gamble   (Fig. 30)

(named after Siam, the old name for Thailand)

Gamble in Ann. Roy. Bot Gard. Calc. 7: 59 (1896); Holttum in Gard. Bull. Sing. 16: 80 (1958).

Densely tufed bamboo; culms erect or erect with arching tips, about 8 m tall, diameter 4-5 cm, internodes with very thick walls and a small lumen, 30 cm long, smooth, green, nodes not swollen; young shoots pale to purplish green. Branch complement of many branches at each node with the middle branch dominant, borne from midculm nodes upwards. Culm sheaths pale to purplish green, becoming stramineous and thin with age, persistent, about 20 cm long, 8-10 cm wide near the base, narrowing to the apex, up to 2.5 cm wide, with pale scattered appressed hairs; blades erect, narrowly lanceolate, 6-15 cm long, 5-12 mm wide near the base, pubescent adaxially; auricles not present; ligule very short, shortly laciniate. Leaf blades 7-14 x 0.5-0.8 cm, pale green, usually glabrous; sheaths hairy along the margins; auricles not present; ligule very short. Inflorescences borne terminally on leafy or leafless branches.

Distribution. Native to Thailand and neighbouring countries, introduced elsewhere in the tropics as an ornamental, because of its very elegant habit.

Habitat. In Thailand *T. siamensis* is found growing naturally in mixed or teak forest; it is also planted on a wide range of soils in a row as a wind break, or in plantations or as an ornamental plant.

Uses. In Thailand the young shoots are consumed as a vegetable like those of other bamboos, and regarded as one of the best bamboo shoots.

Notes. Though not very commonly seen, this species deserves to be planted more widely. Some clumps of this have been planted at the Golf Club in Kota Kinabalu and around Beaufort.

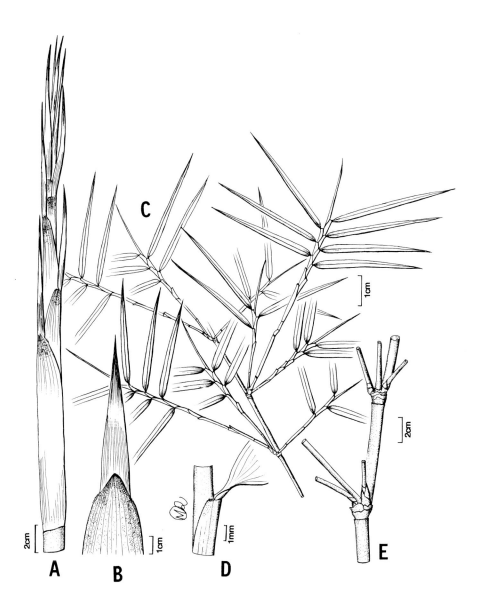

Fig. 30. *Thyrsostachys siamensis*. A, Culm shoot. B, Culm sheath, top part. C, Leafy branches. D, Leaf sheath, detail. E, Mid-culm branch complements. (All from fresh material.)

YUSHANIA Keng f.

(named after Yushan, a mountain in Taiwan, where the genus was first known)

Keng f. in Acta Phytotax. 6: 355 (1957).

     Sympodial bamboo with short to elongated rhizome necks; culms erect, nodes not swollen, supranodal ridge often prominent with few to several spinelike aerial roots below it. Branches at each node 2-several, usually borne on the same plane. Inflorescence paniculate, semelauctant, borne terminally on leafy branches. Spikelets consisting of 4-10 florets; lodicules three; stamens three.

Yushania tessellata (Holttum) S. Dransf. (Fig. 31)

(Latin: *tessellatus*; leaf blades tessellate)

Dransfield in Kew Bull. 37(4): 678 (1983).

Basionym:
Racemobambos tessellata Holtt. in Gard. Bull. Sing. 26: 211 (1973).

     Open-tufted sympodial bamboo, with 5-6 cm spacing between the culms; culms about 2-3(-7) m tall, diameter about 1-1.7 cm, with relatively thin walls, internodes 30-40 cm long, smooth, glabrous, light green to yellowish green, supra-nodal ridge prominent often with few to several spinelike aerial roots below it, nodes not swollen; young shoots light green. Branch complement of 2-several (usually 5) branches borne on the same plane, on or just below the supranodal ridge, developing intravaginally. Culm sheaths smooth, with scattered appressed white hairs when young, becoming glabrous later, 9-18 cm long, 5-6 cm wide near the base, narrowing to the apex, about 5 mm wide, light green turning stramineous with age; blades erect or deflexed, narrowly lanceolate, 0.7-6 cm long, about 3 mm wide, glabrous or minutely hairy; auricle not present, replaced by a few bristles; ligule up to 3 mm long, irregularly toothed. Leaf blades 10-16 x 1-2 cm, with conspicuous cross-veins, base attenuate, usually glabrous; sheaths glabrous; ligule short; auricles not prominent, but with long bristles. Inflorescence a panicle, borne terminally on leafy branches, up to 15 cm long, semelauctant, axes glabrous. Spikelets about 2-5 cm long, pedicels 2-5 mm

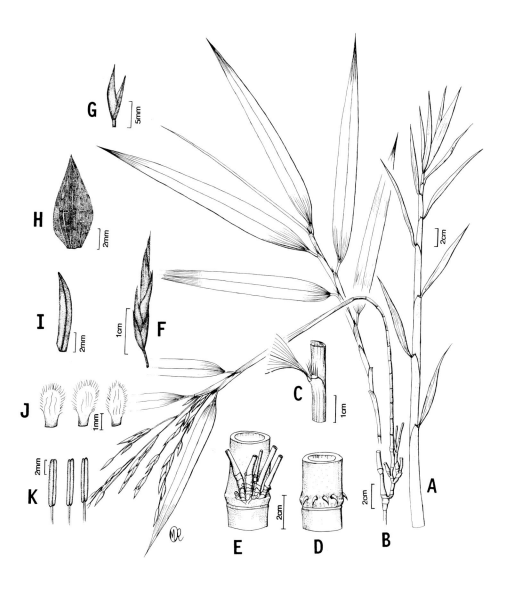

Fig. 31. *Yushania tessellata*. A, Culm shoot. B, Leafy and flowering branches. C, Leaf sheath, detail. D, Basal culm node with root-thorns. E, Mid-culm branch complement. F, Spikelet. G, Glumes. H, Lemma. I, Palea. J, Lodicules. K, Stamens. (All from SD 756.)

long, consisting of 2 glumes, up to 10 florets and a reduced terminal floret, rachilla internodes about 5 mm long; glumes with acuminate tips; lemmas about 5 mm long, usually glabrous, margins hairy; paleas 2-keeled, keels hairy, shorter than the lemma; lodicules 3; stamens 3.

Distribution. So far this bamboo has been found only in Sabah, on Mt. Kinabalu at above 1800 m, on Gunung Alab at 1500 m, and in the Meligan Forest Reserve at 1500 m.

Habitat. Mountain forest.

Notes. The genus *Yushania* was described by Keng in 1957, based on a bamboo found in Taiwan, which is named *Yushania niitakayamensis*. This species is also found in Luzon (Philippines). It has three stamens in the floret. *Y. tessellata* differs from *Y. niitakayamensis* by its attenuate leaf blades and acuminate glumes. The genus *Yushania* has been regarded as synonymous with other genera, such as *Sinarundinaria* Nakai or *Fargesia* Franch. from China, without critical explanation. In this account the name *Yushania* is used to avoid confusion.

"BAMBUSA" sp. (related to B. wrayi Stapf)

This species, known from several montane areas in west Borneo and quite easily observed at Mt. Kinabalu and the Crocker Range in Sabah was preliminarily identified by Holttum (on specimens in the Kew herbarium) as identical with or closely related to *Bambusa wrayi* Stapf from Peninsular Malaysia. Recent studies (K.M. Wong, pers. comm. 1991) have shown that it is distinct from *B. wrayi*, though both species together represent an undescribed genus. Aside from characteristics of the inflorescence and spikelet, this genus differs from *Bambusa* in having a branch complement of many subequal branches where the primary branch is not dominant in size or length. This undescribed genus will be fully treated in a separate study by K.M. Wong. For want of a formal name we refer to the Sabah species as "Bambusa" sp., related to *B. wrayi*.

"Bambusa" sp. related to *B. wrayi* (Fig. 32)

Open-tufted scrambling mountain bamboo; culms 10-20 m long (or more), diameter 1.5-2 cm, the lower part erect, upper part leaning or scrambling over nearby vegetation, or drooping to the ground, with thin walls, internodes 40-80(-120) cm long, green with dark brown hairs when young, becoming smooth and glabrous, nodes (especially those at upper part) swollen. Branch complement of many branches at each node, all of similar size. Culm sheaths about 30 cm long, covered with black hairs when young; blades narrowly lanceolate, hairy adaxially; auricles small, with long bristles; ligule 3 mm tall, laciniate with long bristles (bristles up to 20 mm long). Leafy branches up to 60 cm long, bearing up to 17 leaf blades; leaf blades 11-23 x 1-3 cm, almost sessile, base truncate, glabrous; sheaths glabrous, margins ciliolate; ligule very short, serrate. Inflorescences borne on leafy branches. Spikelets slender, more or less cylindrical, about 2 cm long, rigid, glabrous, consisting of one glume, 3 florets and a reduced terminal floret; glumes glabrous; lemmas glabrous; paleas shorter than lemmas; lodicules 3.

Distribution. Sabah : Mt. Kinabalu, Crocker Range; Sarawak : Kelabit Highlands. Brunei : Pagon Range.

Habitat. Mountain forest.

Vernacular name. Bambas

Uses. The internodes have thin walls, and are used for making musical instruments, such as the *sompoton* (Sabah) and the *engkrui* (Sarawak).

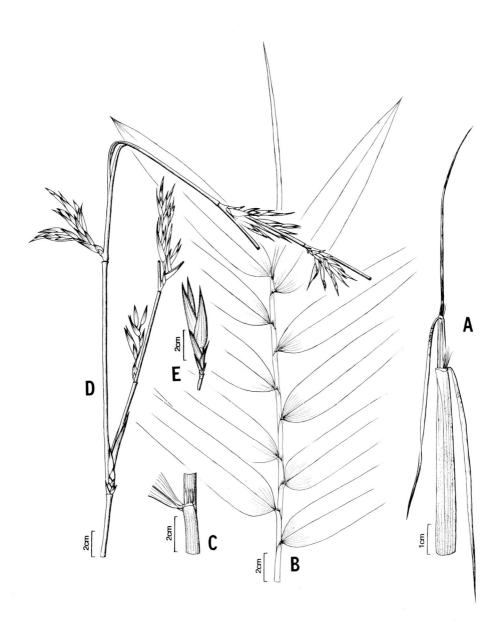

Fig. 32. *"Bambusa"* sp. (related to *B. wrayi*). A, Culm shoot. B, Leafy branch. C, Leaf sheath, detail. D, Flowering branch. E, Pseudospikelet. (A from SD 759; B, C from SD 711; D, E from SD 758.)

REFERENCES

Campbell, J.J.N. (1985) Bamboo flowering patterns: a global view with special reference to East Asia. Journal of the American Bamboo Society 6: 17-35.

Dransfield, S. (1981) The genus *Dinochloa* (Gramineae-Bambusoideae) in Sabah. Kew Bulletin 36(3): 613-633.

Dransfield, S. (1983) The genus *Racemobambos* (Gramineae-Bambusoideae). Kew Bulletin 37(4): 661-679.

Dransfield, S. (1983) Notes on *Schizostachyum* (Gramineae-Bambusoideae) from Borneo and Sumatra. Kew Bulletin 38(2): 321-332.

Dransfield, S. (1989) *Sphaerobambos*, a new genus of bamboo (Gramineae-Bambusoideae) from Malesia. Kew Bulletin 44(3): 425-434.

Dransfield, S. (1989) A new species of *Dinochloa* (Gramineae-Bambusoideae) from Borneo. Kew Bulletin 44(3): 435-437.

Dransfield, S. (1991) Bamboo resources in Thailand: how much do we know? Presented in the IV International Bamboo Workshop, Chiangmai, Thailand, 27-30 November 1991.

Holttum, R.E. (1958) The bamboos of the Malay Peninsula. Gardens' Bulletin, Singapore 16: 1-135.

Kiang, Tao & Wei-Chih Lin (1978) An investigation of bamboo resources in Sabah. Rural Development Corporation, Sabah: Kota Kinabalu, Sabah. Unpublished duplicated report, 24 p.

Kulip, J. (1992) *Racemobambos glabra*, a new record for Sabah, Malaysia. Sandakania 1: 7-9.

Liew, T.C. (1973) Eradication of climbing bamboo in dipterocarp forests of Sabah. Malaysian Forester 36(4): 243-256.

McClure, F.A. (1966) Bamboos. A fresh perspective. Harvard University Press: Cambridge, Massachusetts. 347 p.

Wong, K.M. (1991) *Schizostachyum terminale* Holtt., an interesting new bamboo record for Borneo. Gardens' Bulletin, Singapore 43: 39-42.

Wong, K.M. (1992) The Poring Puzzle: *Gigantochloa levis* and a new species of *Gigantochloa* (Gramineae: Bambusoideae) from Peninsular Malaysia. Sandakania 1: 15-21.

Wong, K.M., C.L. Chan and A. Phillipps (1988) The gregarious flowering of Miss Gibbs' bamboo (*Racemobambos gibbsiae*) and Hepburn's bamboo (*R. hepburnii*) on Mount Kinabalu. Sabah Society Journal 8(4): 466-474.

CHECKLIST OF BAMBOOS IN SABAH

Bambusa

B. blumeana Schult.
B. heterostachya (Gamble) Holtt.
B. multiplex (Lour.) Raeuschel ex J.A. & J.H. Schult.
B. tuldoides Munro
B. vulgaris Schrader ex Wendland
B. sp. (unidentified)

Dendrocalamus

D. asper (Schultes f.) Backer ex Heyne

Dinochloa

D. darvelana S. Dransf.
D. obclavata S. Dransf.
D. prunifera S. Dransf.
D. robusta S.Dransf.
D. scabrida S. Dransf.
D. sipitangensis S. Dransf.
D. sublaevigata S. Dransf.
D. trichogona S. Dransf.
D. sp. (unidentified)

Gigantochloa

G. balui K.M. Wong
G. levis (Blanco) Merrill

Racemobambos

R. gibbsiae (Stapf) Holtt.
R. glabra Holtt.
R. hepburnii S. Dransf.
R. hirsuta Holtt.
R. pairinii K.M. Wong
R. rigidifolia Holtt.

Schizostachyum

S. blumei Nees
S. brachycladum Kurz
S. latifolium Gamble
S. lima (Blanco) Merr.
S. pilosum S. Dransf.
S. terminale Holtt.
S. sp. (unidentified)

Sphaerobambos

S. hirsuta S. Dransf.

Thyrsostachys

T. siamensis Gamble

Yushania

Y. tessellata (Holtt.) S. Dransf.

Unnamed genus

(one species)

INDEX TO SCIENTIFIC NAMES

(bold numbers indicate detailed treatment;
italicized numbers indicate illustrations)

INDEX TO VERNACULAR NAMES

and names in English